Advanced Maths Essentials
Mechanics 1 for AQA

D0505083

1 Mathematical modelling 1
  1.1 Mathematical modelling in mechanics 1

2 Kinematics in one and two dimensions 3
  2.1 Displacement, speed, velocity and acceleration 3
  2.2 Kinematics graphs 3
  2.3 Use of average speed and average velocity 7
  2.4 Constant acceleration equations 8
  2.5 Motion in a vertical plane 11
  2.6 Vectors 15
  2.7 The unit vectors, $\mathbf{i}$ and $\mathbf{j}$ 16
  2.8 Application of vectors 18
  2.9 Equations of motion with vectors 20
  2.10 Velocity triangles and resultant velocities 22

3 Statics and forces 27
  3.1 Resultant of forces 27
  3.2 Equilibrium of forces 29
  3.3 Types of force 32
  3.4 Friction and the coefficient of friction 34

4 Momentum 41
  4.1 Momentum 41
  4.2 The principle of conservation of momentum 42

5 Newton's laws of motion 47
  5.1 Newton's laws of motion and application of $\mathbf{F} = m\mathbf{a}$ 47
  5.2 Friction and $\mathbf{F} = m\mathbf{a}$ 48

6 Connected particles 54
  6.1 Connected particles 54

7 Projectiles 62
  7.1 Projectiles 62
  7.2 Range, time of flight and maximum height 65
  7.3 Modification of the equations of motion 67

Practice exam paper 73

Answers 75

Welcome to Advanced Maths Essentials: Mechanics 1 for AQA. This book will help you to improve your exam performance by focusing on all the essential maths skills you will need in your AQA Mechanics 1 exam. It has been divided by chapter into the main topics that need to be studied. Each chapter has then been divided by sub-headings, and the description below each sub-heading gives the AQA specification for that aspect of the topic.

The book contains scores of worked examples, each with clearly set-out steps to help solve the problem. You can then apply the steps to solve the Skills Check questions in the book and past exam questions at the end of each chapter. If you feel you need extra practice on any topic, you can try the Skills Check Extra exercises on the accompanying CD-ROM. At the back of this book there is a practice exam-style paper to help you test yourself before the big day.

Some of the questions in the book have a 	 symbol next to them. These questions have a PowerPoint® solution (on the CD-ROM) that guides you through suggested steps in solving the problem and setting out your answer clearly.

Using the CD-ROM

To use the accompanying CD-ROM simply put the disc in your CD-ROM drive, and the menu should appear automatically. If it doesn't automatically run on your PC:

1. Select the My Computer icon on your desktop.
2. Select the CD-ROM drive icon.
3. Select Open.
4. Select mechanics1_for _aqa.exe

If you don't have PowerPoint® on your computer you can download PowerPoint 2003 Viewer®. This will allow you to view and print the presentations. Download the viewer from http://www.microsoft.com

Pearson Education Limited
Edinburgh Gate
Harlow
Essex
CM20 2JE
England
www.longman.co.uk

First published 2005
Fifth impression 2011
ISBN   978-0-582-83689-1

Design by Ken Vail Graphic Design

Cover design by Raven Design

Typeset by Tech-Set, Gateshead

Printed in China EPC/05

The Publisher wishes to draw attention to the Single-User Licence Agreement situated at the back of the book. Please read this agreement carefully before installing and using the CD-ROM.

We are grateful for permission from the Assessment and Qualifications Alliance to reproduce past exam questions. All such questions have a reference in the margin. The Assessment and Qualifications Alliance can accept no responsibility whatsoever for accuracy of any solutions or answers to these questions.

Every effort has been made to ensure that the structure and level of sample question papers matches the current specification requirements and that solutions are accurate. However, the publisher can accept no responsibility whatsoever for accuracy of any solutions or answers to these questions. Any such solutions or answers may not necessarily constitute all possible solutions.

1 Mathematical modelling

1.1 Mathematical modelling in mechanics

Use of assumptions; mathematical analysis; interpretation and validity; refinement and extension.

Mathematical models can be used in mechanics to make predictions about real-life situations. Models must be tested and improved by comparing their predictions against experimental data. The models can then be refined or extended to describe the situation more accurately.

In order to develop a model you must simplify the real-life situation by making **modelling assumptions.** For example, if you are trying to model a car that travels 5 km, you may ask where you treat as the starting point of measurement. Do you measure from the front end of the car or the back end? Because the car is small relative to the distance travelled, you may assume that the car can be treated as a particle. This then becomes a modelling assumption. Similarly, the time taken for the car to travel may be influenced by wind; it will reach a given destination faster if there is a strong wind travelling in its direction of travel. In M1 you assume that any effects due to wind can be neglected. There are several terms and definitions used in M1 that you should know (you may be examined on these types of modelling assumptions):

Objects treated as particles: in general, it is assumed that objects that are small relative to the other sizes involved will be treated as particles; this means that the mass of the object can be considered to act at the single point where the particle is placed, e.g. an aeroplane can be modelled as a point particle, as it travels long distances compared to its length.

Lamina: this is a flat object whose thickness can be ignored, as the thickness is small compared to the other dimensions (its width and its length). For example, a thin sheet of card may be represented as a lamina.

Rigid body: this is an object that is made up of different parts or masses which do not move relative to each other. For example, a car can be considered to be a rigid body, i.e. when the engine produces a forward force, the whole car moves.

Rod: this is an object in which the mass is considered to concentrate along a line; it is assumed to have only length, and not breadth nor width.

Uniform object (lamina or rod): if an object is uniform then the mass is evenly distributed across the object and can be considered to act at its centre. For example, you could make the assumption that a tree trunk is a uniform rod and that its mass can be considered to act from its centre. However, this may not be a reasonable modelling assumption if the base of the trunk is thicker than the top of the trunk. In this case, the trunk is non-uniform and the centre of mass may have to be calculated.

Light object: when the mass of an object is small compared to the rest of the masses involved, it is said to be a light object. For example, when modelling the mass of the Earth, the mass of a car is relatively light, so the mass of the car may be ignored in the model.

Light and inextensible strings: in M1, strings are assumed to be:

1 light, i.e. the mass of the string is small relative to the other masses involved. This is a fair assumption, unless you are using a chain instead of a string, in which case the mass may have to be included.

2 inextensible, i.e. you don't have to account for any force to extend the string. Again, this is a fair assumption, as the extension in a string is relatively small.

More complicated problems, without these assumptions, are introduced in M3.

Smooth surface: when a surface is assumed to be smooth then you assume that there is no resistive force opposing the movement due to the contact of the surface with the object, i.e. you assume that there is no **friction**. For example, when modelling an ice rink, you may assume that the surface is smooth.

Note:
The experimental laws of friction are covered in more detail in Chapter 3.

Rough surface: when the surface is not smooth then it is said to be rough. In this case the friction must be included when modelling the situation. For example, sand paper being rubbed across a brick will have a comparatively large force of friction that cannot be ignored because the contact between the surfaces is rough.

Smooth and light pulleys: in M1, pulleys are assumed to be:

1 smooth, i.e. there is no friction on the surface or the bearings of the pulley.

2 light, i.e. the mass of the pulley is considered to be small in comparison to the other masses involved.

Bead: this is a particle that can be threaded onto a string or wire.

Wire: this is a rigid body that is a thin thread of metal.

Peg: this is a support from which an object can be hung or on which an object may rest. It acts as a point but may be either rough or smooth.

Gravity: this is the force that attracts objects towards each other. Because of the relatively massive size of the Earth, it can be assumed that objects only experience attraction towards the Earth and not towards each other. The force of gravity reduces in size as you move further away from the surface of the Earth. However, in all the models in M1, gravity is assumed to be constant.

Air resistance: this is a force that is experienced because of the resistance of the air. For example, when you drop a sheet of paper, the paper will not fall as fast as a brick. However, in M1 it is generally assumed that there is no air resistance.

Wind: this is a force that can be felt because of the action of the wind. In M1 it is assumed that there is no wind.

2 Kinematics in one and two dimensions

2.1 Displacement, speed, velocity and acceleration

Understanding the relationship between displacement, speed, velocity and acceleration.

The following quantities are needed to describe the motion of a particle.

Distance is the length of a given path. We measure this in metres, written m.

Displacement defines the position of one point relative to another point: displacement includes both the distance between the two points and the direction of the first point from the second. Displacement is measured in metres, written m.

Speed is the rate of change of the distance with time, with no account taken for direction. The units for speed are metres per second, written m/s or m s^{-1}.

Velocity is the rate of change of displacement with time, and so the direction is taken into account. The units for velocity are metres per second, written m/s or m s^{-1}.

The **acceleration** is the rate of change of velocity with time. The units for acceleration are metres per second squared, written m/s^2 or m s^{-2}.

2.2 Kinematics graphs

Sketching and interpreting kinematics graphs.

The aim of kinematics is to analyse the motion of a particle that is travelling in a straight line – its velocity, time of travel, acceleration and the path that it follows. In this unit, the particle always has a constant acceleration.

The properties that you are analysing are (with standard units in brackets):
s = displacement of particle from a fixed point (metres, m)
u = the initial velocity of the particle (m/s or m s^{-1})
v = the final velocity of the particle (m/s or m s^{-1})
a = the acceleration of the particle (m/s^2 or m s^{-2})
t = time taken for the particle to travel (seconds, s).

A **speed–time graph** shows how the speed of a particle varies with time. The horizontal axis represents the time taken and the vertical axis represents the speed. In the graph, u is the initial velocity and v is the final velocity after travelling for t seconds under constant acceleration.

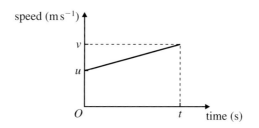

> **Recall:**
> Acceleration is the rate of change of velocity of a particle, where velocity is the speed of a particle in a given direction.

> **Note:**
> These variables are often referred to as *suvat*.

You can obtain important information from a speed–time graph.

The gradient of the line $= \dfrac{(v - u)}{t} =$ change of speed over time, a.

The area under the line $= \frac{1}{2}(u + v)t =$ average speed $\times$ time $=$ distance travelled.

Remember, in a speed–time graph:
  the gradient of the line = the acceleration of the particle
  the area under the line = the distance travelled by the particle

A **velocity–time** graph shows how the velocity of a particle varies with time. In this case the area represents the displacement, and *when the area is below the time axis this represents a negative displacement.*

A **displacement–time** graph shows how the displacement of a particle varies with time. The horizontal axis represents the time taken and the vertical axis represents the displacement. We also know that:

$$\text{velocity} = \frac{\text{change in displacement}}{\text{time taken}}, \text{ and so:}$$

  the gradient of a displacement–time graph
  = the velocity of the particle

A particle that slows down is **decelerating**. When this is the case, the acceleration is negative; this can also be called retardation.

Recall:
From C1, the gradient, m, between two points (x_1, y_1) and (x_2, y_2) is:
$$m = \frac{y_2 - y_1}{x_2 - x_1}$$

Recall:
The area of a trapezium is $\frac{1}{2}(a + b)h$, where a and b are the lengths of the parallel sides and h is the height.

Note:
Distance $=$ (average) speed $\times$ time.

Note:
Acceleration is the rate of change
of velocity $= \dfrac{\text{change in velocity}}{\text{time}}$
when the acceleration is constant.

Example 2.1 A particle travels with constant speed 4 m s^{-1} from A to B for 4 seconds. It then turns around and travels in the opposite direction from B to A with constant speed 2 m s^{-1} for a further 2 seconds.

a Sketch a velocity–time graph to represent the journey of the particle in the first 6 seconds.

b Calculate the displacement **i** after the first 4 seconds and **ii** after 6 seconds of the motion.

c Hence, sketch a displacement–time graph.

Step 1: Draw a clear diagram to represent the information given.

a

velocity (m s^{-1})

[graph: velocity 4 from 0 to 4 s, then −2 from 4 to 6 s]

Step 2: Use area = displacement to solve the problem.

b i Area under the graph for the first 4 seconds $= 4 \times 4 = 16$.
So the displacement after the first 4 seconds is 16 m.

ii Area 'under' the graph for the last 2 seconds $= 2 \times 2 = 4$.
So the displacement after 6 seconds is $16 + (-4) = 12$ m.

Recall:
The gradient must be constant for the displacement–time graph because the velocity is constant.

c

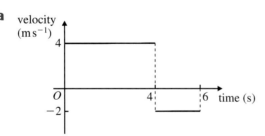

displacement (m)

[graph: displacement rising from 0 to 16 at time 4, then falling to 12 at time 6]

Example 2.2 A car starts from a point *A* and accelerates uniformly at 1 m s^{-2} from rest for three seconds. It then maintains a steady speed for a further 15 seconds. Finally it slows down for 5 seconds with uniform retardation until it stops at point *B*.

   a Sketch a speed–time graph for the journey of the car.

   b Find the maximum speed of the car.

   c Find the acceleration of the car in the final stage of the journey.

   d Find the distance *AB*.

Step 1: Draw a clear diagram to represent the information given.

   a Let the maximum speed be $v \text{ m s}^{-1}$.

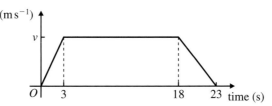

Step 2: Use gradient = acceleration and area = distance to solve the problem.

   b Acceleration in first 3 seconds = gradient in first three seconds

$$1 = \frac{(v - 0)}{(3 - 0)}$$

$$v = 3$$

The maximum speed of the car is 3 m s^{-1}.

   c Acceleration in last 5 seconds = gradient in last five seconds

$$a = \frac{0 - v}{5} = -\frac{3}{5}$$

The acceleration of the car in the last 5 seconds is -0.6 m s^{-2}.

   d Distance travelled from *A* to *B* = total area

$$\text{area of trapezium} = \tfrac{1}{2}(a + b)h$$

$$= \tfrac{1}{2} \times (23 + 15) \times 3$$

$$= 57$$

The distance *AB* is 57 m.

Note:
You could also use the equations of motion, studied in Section 2.4, to solve this problem.

Note:
The area could also be calculated by dividing the trapezium into two triangles and a rectangle, and then summing their areas.

An **acceleration–time** graph shows how the acceleration of a particle varies with time. This can be sketched from a velocity–time graph by using the fact that the gradient of a velocity–time graph gives the acceleration.

Example 2.3 At $t = 0$, where *t* is the time in seconds, a train is travelling with a uniform velocity of 7 m s^{-1}. The train then approaches a station and, at $t = 4$, the driver applies the brake slowing down the train with uniform retardation. It reaches the station *T* seconds after applying the brakes, where it comes to a halt.

   a Sketch a velocity–time graph of the journey of the train.

   b Calculate the retardation, leaving your answer in terms of *T*.

The distance travelled by the train during the first 4 seconds is three quarters of the distance travelled during *T* seconds.

   c Calculate the value of *T* and the retardation.

   d Hence sketch an acceleration–time graph of the journey of the train.

a

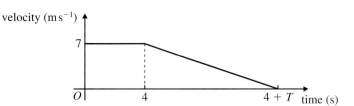

b Acceleration in last T seconds = gradient in last T seconds

$$a = \frac{(v - u)}{t} = \frac{(0 - 7)}{T} = -\frac{7}{T}$$

Hence, the retardation of the train in the last T seconds is $\frac{7}{T}$ m s^{-2}.

> **Recall:**
> Acceleration is given by the gradient in a speed–time graph.

> **Recall:**
> Distance travelled is given by the area under a speed–time graph.

c Distance travelled in first 4 s = 0.75 × (distance travelled in last T s)

$$4 \times 7 = 0.75 \left(\frac{1}{2} \times T \times 7 \right)$$

$$T = \frac{32}{3}$$

The train slows for $10\frac{2}{3}$ s.

$$\text{Retardation} = \frac{7}{T} = \frac{21}{32} = 0.65625$$

The retardation of the train is 0.656 m s^{-2} (3 s.f.).

> **Note:**
> Substitute this value for T into the relation in **b**.

d Acceleration–time graph:

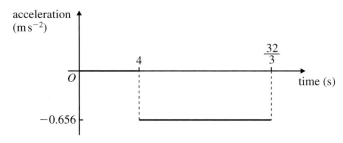

Example 2.4 A motorbike tries to catch up with a car. The car leaves 20 seconds before the motorbike, and travels with a constant speed of 8 m s^{-1}. The motorbike accelerates uniformly at 2 m s^{-2}, until it reaches a speed of 12 m s^{-1}. It then maintains this constant speed for the remainder of the journey. Find the time taken for the motorbike to reach the car.

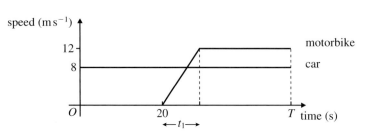

> **Note:**
> The car leaves at $t = 0$, the motorbike leaves at $t = 20$ and the two meet at $t = T$.

Acceleration of motorbike = gradient of speed–time graph during t_1 s.

$$2 = \frac{12}{t_1}$$

$$t_1 = 6$$

When the motorbike reaches the car

Area under line (car) = area under line (motorbike)

$$8 \times T = \tfrac{1}{2} \times 12 \times ((T - 20) + (T - 26))$$

$$T = 69$$

The motorbike catches up with the car 69 seconds after the car left.

Hence it takes the motorbike $(69 - 20)$ s = 49 s to reach the car.

2.3 Use of average speed and average velocity

Average speed and average velocity.

The **average speed** of a given journey is the constant speed that would be required to cover the distance of the journey in the given time:

average speed in a given interval of time =

$$\frac{\text{distance covered in the given interval of time}}{\text{interval of time}}$$

The **average velocity** for a given journey is the uniform velocity that would be required to achieve the change in displacement from the start of the journey to the end of the journey:

average velocity in a given interval of time =

$$\frac{\text{change in displacement in the given interval of time}}{\text{interval of time}}$$

In a displacement–time graph the gradient of the line joining the start point to the end point represents the average velocity.

Example 2.5 A particle starts from a point O and travels 50 m due north in 10 seconds to a point A. The particle then travels 70 m due south in 5 seconds to a point B.

  a Sketch a displacement–time graph of the journey of the particle from O to B.

  b Find the average velocity of the particle in its journey from O to B.

a

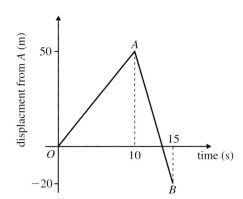

b The total displacement from O to B is -20 m.

$$\text{Average velocity from } O \text{ to } B = \frac{\text{change in displacement from } O \text{ to } B}{\text{interval of time}}$$

$$= \frac{-20}{15}\, \text{m s}^{-1}$$

$$= -\frac{4}{3}\, \text{m s}^{-1}$$

Note:
This value is the same as the gradient of the straight line from O to B.

2.4 Constant acceleration equations

Use of constant acceleration equations..

When the acceleration is constant or uniform, then the variables s, u, v, a and t (referred to as *suvat*) can be related using equations.

These **equations of motion** are called the **constant acceleration equations.** They are:

$s = ut + \frac{1}{2}at^2$

$s = vt - \frac{1}{2}at^2$

$v = u + at$

$s = \frac{1}{2}(u + v)t$

$v^2 = u^2 + 2as$

Note:
You should learn these formulae.

Note:
The units must be consistent when using these equations.

Each equation contains four variables. In most questions you will be given three variables and asked to calculate the fourth.

Example 2.6 A particle moves in a straight line from A to B. The particle starts from rest at A and accelerates at 2 m s^{-2} until it reaches a speed of 8 m s^{-1} at B.

a Find how long it takes to travel from A to B.

b Find the distance AB.

Tip:
If a particle starts from rest, this is another way of saying $u = 0$.

Step 1: Draw a clear diagram to represent the information given.

$0\,\text{m s}^{-1}$ $2\,\text{m s}^{-2}$ $8\,\text{m s}^{-1}$
A ———————————————————— B

Note:
Acceleration is represented by a double arrow ⟶ and velocity by a single arrow →.

Step 2: Fill the information in *suvat*, identifying what is required with a question mark.

a s not required   **b** $s = ?$

$u = 0$ $u = 0$

$v = 8$ $v = 8$

$a = 2$ $a = 2$

$t = ?$ t not required

Step 3: Pick an equation of motion relating the three known variables with the unknown that is required, insert values, rearrange (if necessary) and solve.

$v = u + at$ $v^2 = u^2 + 2as$

$8 = 0 + 2t$ $8^2 = 0^2 + 2(2)s$

$t = 4$ $s = 16$

It takes 4 s to travel from A to B.   Distance AB is 16 m.

Tip:
Don't forget the units.

A particle that slows down is **decelerating**, in which case the acceleration is negative. If the acceleration of a particle is -7 m s^{-2}, then its **deceleration** or **retardation** is 7 m s^{-2}.

Tip:
You could use the value $t = 4$ found in part **a** to calculate s but it is bette to use the values given in the question in case your answer to **a** is incorrect.

Example 2.7 A train is travelling at $60\ \text{km h}^{-1}$ on a straight railway track. The driver of the train sees a red signal 500 m ahead. He immediately applies the brakes so that the train decelerates at $0.25\ \text{m s}^{-2}$.

   a Find how far the train is past the signal when it comes to a halt.

   b What is the velocity of the train as it passes the signal?

Step 1: Draw a clear diagram to represent the information given.

$60\ \text{km h}^{-1}$ $-0.25\ \text{m s}^{-2}$ signal

← 500 m →

Step 2: Fill the information in *suvat*, identifying what is required with a question mark.

a Motion of train from the start until it stops:

$s = ?$
$u = 16\frac{2}{3}$
$v = 0$
$a = -0.25$
t not required

b Motion of train from the start until it passes the signal:

$s = 500$
$u = 16\frac{2}{3}$
$v = ?$
$a = -0.25$
t not required

Step 3: Pick an equation of motion relating the three known variables with the unknown that is required, insert values, rearrange (if necessary) and solve.

$v^2 = u^2 + 2as$
$(0)^2 = (16\frac{2}{3})^2 + 2(-0.25)s$
$s = 556 \text{ (3 s.f.)}$

The train stops $(556 - 500)$ m $= 56$ m after the signal.

$v^2 = u^2 + 2as$
$v^2 = (16\frac{2}{3})^2 + 2(-0.25)500$
$v = 5.27 \text{ (3 s.f.)}$

The train travels at $5.27\ \text{m s}^{-1}$ as it passes the red signal.

> **Note:**
> You must work in consistent units.
> $1000\ \text{m} = 1\ \text{km}$ and $3600\ \text{s} = 1$ hour, so to convert from km h^{-1} to m s^{-1} you multiply by 1000 and divide by 3600. In this case, $60\ \text{km h}^{-1} = 16\frac{2}{3}\ \text{m s}^{-1}$

> **Note:**
> Be careful not to confuse the variable *s* (displacement) with the unit s (seconds).

> **Note:**
> Only the positive root is physically reasonable.

Example 2.8 Two cars, A and B, start from the same point O and travel in a straight line. Car A leaves at $t = 0$ seconds and travels with constant speed $5\ \text{m s}^{-1}$. Car B leaves 10 seconds later and accelerates uniformly at $5\ \text{m s}^{-2}$. Find the distance from O when car B overtakes car A.

Suppose they meet T seconds after car A leaves O. t_A and t_B are the times for each car, measured from the start of A's journey:

Step 1: Draw a clear diagram to represent the information given.

For A:
$t_A = 0$
$u = 5$ $a = 0$ $t_A = T$
O

For B:
$t_B = 10$
$u = 0$ $a = 5$ $t_B = T$
O

Step 2: Fill the information in *suvat*, identifying what is required with a question mark.

For car A:
$s = ?$
$u = 5$
v not required
$a = 0$
$t = T$

For car B:
$s = ?$
$u = 0$
v not required
$a = 5$
$t = T - 10$

Step 3: Pick an equation of motion relating the three known variables with the unknown that is required, insert values, rearrange (if necessary) and solve.

$s = ut + \frac{1}{2}at^2$
$s = 5T + \frac{1}{2}(0)T^2$
$= 5T$

$s = ut + \frac{1}{2}at^2$
$s = (0)T + \frac{1}{2}(5)(T-10)^2$
$= 2.5T^2 - 50T + 250$

When B overtakes A
$5T = 2.5T^2 - 50T + 250$
$0 = 2.5T^2 - 55T + 250$
$T = 15.58\ldots \text{ or } 6.42\ldots$

$T = 15.58\ldots$ is the time that they meet because $T = 6.42\ldots$ would give a negative time t for car B.
The cars have travelled $5 \times 15.58\ldots = 77.9$ m (3 s.f.) when they meet.

> **Note:**
> A particle travels with constant speed is another way of saying $a = 0$.

> **Note:**
> Consider the journey to the point where both cars have travelled the same distance.

> **Note:**
> When there are two motions, apply *suvat* to each situation and find the common link. In this case, if car A has travelled for T s then car B has travelled for $(T - 10)$ s.

> **Note:**
> Pick equations that include both *s* and *t*.

> **Recall:**
> Using the quadratic formula from C1.

Example 2.9 A particle moves in a straight line. It passes a point O with velocity u m s^{-1} and has constant acceleration a m s^{-2}. Two seconds later it passes a point P. One second after it passes the point P it passes a point Q. Given that the distance OP is 34 m and the distance PQ is 20 m, find u and a.

Step 1: Draw a clear diagram to represent the information given.

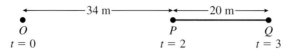

Step 2: Fill the information in *suvat*, identifying what is required with a question mark.

Motion from O to P:	Motion from O to Q:
$s = 34$	$s = 54$
$u = ?$	$u = ?$
v not required	v not required
$a = ?$	$a = ?$
$t = 2$	$t = 3$

> **Note:**
> Analyse the motion from O to P and from O to Q because both sets of equations will have the same two unknowns, u and a.

Step 3: Pick an equation of motion relating the three known variables with the unknown that is required, insert values, rearrange (if necessary) and solve.

Motion from O to P:

$$s = ut + \tfrac{1}{2}at^2$$
$$34 = 2u + 2a$$
$$17 = u + a \qquad ①$$

Motion from O to Q:

$$s = ut + \tfrac{1}{2}at^2$$
$$54 = 3u + 4.5a$$
$$18 = u + 1.5a \qquad ②$$

> **Note:**
> In this case there are two equations and two unknowns: solve them simultaneously.

Solving ① and ② simultaneously gives $u = 15$ and $a = 2$.

SKILLS CHECK **2A: Kinematics graphs and motion in a horizontal line**

1 A car leaves a point O and accelerates uniformly from rest to a speed of 4 m s^{-1} in 2 seconds. It then maintains a steady speed for a further 20 seconds, after which it slows down to a halt in 5 seconds.

  a Sketch a speed–time graph for the car's journey.

  b Find the acceleration during the initial and final stages of the journey.

  c Find the total distance travelled by the car.

2 A train travels between two stations, P and Q. The train accelerates uniformly from rest to 6 m s^{-1} in 5 seconds. It then continues its journey for a further 15 seconds with this velocity, before decelerating uniformly to a halt at station Q, in a further T seconds. The distance travelled in the last T seconds is $\tfrac{1}{6}$ of the total distance travelled.

  a Sketch a velocity–time graph of the motion of the train between the two stations.

  b Find T.

  c Find the distance PQ.

  d By first finding the acceleration during each stage of the journey, sketch an acceleration–time graph for the journey.

  e Find the average speed of the train's journey from P to Q.

3 Two cars set off on a journey. The first car leaves at time $t = 0$, where t is measured in seconds. It accelerates uniformly until it reaches a speed of 4 m s^{-1} at $t = 3$. It then maintains a constant velocity. The second car leaves from the same point at $t = 3$ and travels with constant speed 8 m s^{-1}.

  a On the same axes sketch speed–time graphs for the motion of the two cars.

  b Find t when the two cars meet.

  c How far are they from the start at this time?

  d Sketch a displacement–time graph for the journey of the second car.

4 A particle travels in a straight horizontal line. It accelerates uniformly from rest to a speed of 4 m s^{-1} in 2 seconds.

   a Find the acceleration.   **b** Find the distance travelled by the particle.

5 An object undergoing uniform acceleration travels 100 m in 12 seconds. Given the initial velocity is zero, find the acceleration.

6 A particle moves along a straight line AB with constant acceleration of 1 m s^{-2}. If AB is 20 m and it takes 2 seconds to travel from A to B, what was the velocity of the particle at A?

7 A car travels in a straight line with uniform acceleration of 3 m s^{-2}.

   a If the initial velocity is 8 m s^{-1}, how long will it take to travel 6 m?

   b What assumptions have you made in modelling this situation?

8 A man drives a car with a constant acceleration of 5 m s^{-2}. After 2 seconds he sees a set of traffic lights and slows down with a retardation of 2 m s^{-2}. Given that his initial velocity was 2 m s^{-1} and that he manages to stop at the traffic lights, find the distance between the traffic lights and the point where he begins to accelerate.

9 A jet starts from rest and travels in a straight line with a constant acceleration of 10 m s^{-2}.

   a How long will it take to reach a speed of 500 km h^{-1}?

   b How far will it travel in this time?

10 A particle moves in a straight line with constant acceleration. At $t = 0$, $t = 4$ and $t = 8$, where t is the time measured in seconds, the particle passes points P, Q and R respectively. The distance PQ is 100 m and the velocity at P is 20 m s^{-1}.

   a Find the acceleration.   **b** Find the distance QR.

 11 A cyclist leaves a point O from rest and accelerates at 1 m s^{-2}. Three seconds later a motorist, wishing to catch the cyclist, leaves from rest from the same point O and accelerates at 8 m s^{-2}.

   a How far apart are the car and the bicycle 4 seconds after the cyclist left O?

   b After how long do they meet?

   c How far are they from O when they meet?

SKILLS CHECK **2A EXTRA** is on the CD

2.5 Motion in a vertical plane

Vertical motion under gravity.

Any particle travelling vertically experiences a constant acceleration towards the Earth due to gravity of magnitude 9.8 m s^{-2}. Gravity can be assumed to be constant (this is a fair modelling assumption), so the equations of motion can be applied.

Recall:
Modelling assumptions in Chapter 1.

Taking downwards as positive, the particle shown has a velocity of -6 m s^{-1} and an acceleration of 9.8 m s^{-2}.

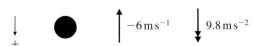

Note:
You can define either up or down as positive but be consistent within a problem.

Taking upwards as positive, the particle shown has a velocity of 6 m s^{-1} and an acceleration of -9.8 m s^{-2}.

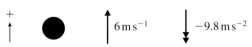

Example 2.10 A ball is thrown upwards with speed $18\ \mathrm{m\ s^{-1}}$.

    a Find the displacement of the particle after 1 second.

    b Find the velocity after **i** 1 second and **ii** 2 seconds.

Step 1: Draw a clear diagram to represent the information given.

Taking upwards as positive:

Recall:
Displacement is always measured from the starting point.

Note:
Decide at the start which direction (up or down) is positive. In this case, up is positive.

Step 2: Fill the information in *suvat*, identifying what is required with a question mark.

a $s = ?$

    $u = 18$

    v not required

    $a = -9.8$

    $t = 1$

Step 3: Pick an equation of motion relating the three known variables with the unknown that is required, insert values, rearrange (if necessary) and solve.

    $s = ut + \frac{1}{2}at^2$

    $s = (18 \times 1) + \frac{1}{2}(-9.8) \times 1^2$

      $= 13.1$

After 1 second the particle has a displacement of 13 m (2 s.f.).

b s not required

    $u = 18$

    $v = ?$

    $a = -9.8$

    $t = 1$ and 2

    $v = u + at$

When $t = 1$

    $v = 18 - (9.8 \times 1)$

      $= 8.2$

i   After 1 second the particle has a velocity of $8.2\ \mathrm{m\ s^{-1}}$.

When $t = 2$

    $v = 18 - (9.8 \times 2) = -1.6$

ii   After 2 seconds the particle has a velocity of $-1.6\ \mathrm{m\ s^{-1}}$.

Note:
Quote solutions to 2 s.f. when using $g = 9.8\ \mathrm{m\ s^{-1}}$, unless otherwise stated.

Note:
The negative sign indicates that the particle is travelling downwards.

Example 2.11 A particle is thrown downwards in the air and moves freely under gravity. It reaches twice its initial velocity after 2 seconds.

    a Find the initial velocity of the particle.

    b Find the displacement of the particle after two seconds.

Taking down as positive:

Note:
In this example, it is better to define down as positive because there is no motion upwards.

Step 1: Draw a clear diagram to represent the information given.

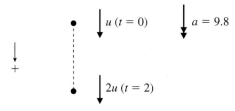

Step 2: Fill the information in *suvat*, identifying what is required with a question mark.

a s not required

    $u = ?$

    $v = 2u$

    $a = 9.8$

    $t = 2$

Step 3: Pick an equation of motion relating the three known variables with the unknown that is required, insert values, rearrange (if necessary) and solve.

    $v = u + at$

    $2u = u + 9.8 \times 2$

    $u = 19.6$

The initial velocity of the particle is $20\ \mathrm{m\ s^{-1}}$ (2 s.f.).

b $s = ?$

    $u = 19.6$

    $v = 39.2$

    $a = 9.8$

    $t = 2$

    $s = \frac{(u + v)}{2}t$

    $s = \frac{(19.6 + 39.2)}{2} \times 2$

      $= 58.8$

The displacement of the particle after 2 seconds is 59 m (2 s.f.).

The **maximum height** of a particle when it is thrown upwards is the height at which it stops travelling up and starts falling down. At this height its velocity in the vertical direction is 0 m s^{-1}. This is an important condition when calculating maximum height reached.

Example 2.12 A cricket ball is thrown upwards at a speed of 14 m s^{-1}. By modelling the ball as a particle, find **a** the maximum height reached and **b** the total distance travelled when it has come back to its starting point.

Taking up as positive:

Step 1: Draw a clear diagram to represent the information given.

$v = 0$

$+$

-9.8 m s^{-2}

$u = 14$

Step 2: Fill the information in *suvat*, identifying what is required with a question mark.

$s = ?$
$u = 14$
$v = 0$
$a = -9.8$
t not required

Step 3: Pick an equation of motion relating the three known variables with the unknown that is required, insert values, rearrange (if necessary) and solve.

$v^2 = u^2 + 2as$
$0^2 = 14^2 + 2(-9.8)s$
$s = 10$

a The maximum height reached is 10 m.

b The total distance travelled is $2 \times 10 = 20$ m.

Note:
Using $v = 0$ at the maximum height defines the third piece of information required to fill *suvat*.

Note:
From the start to the end of the motion the displacement is 0 m. Displacement is measured from the starting point.

The **time of flight** is the total time that a particle is in the air.

Example 2.13 A girl throws a ball vertically upwards in the air with speed 15 m s^{-1} and the ball travels freely under gravity. Find:

a the time of flight of the ball, assuming that it is thrown from the ground

b the time for which the ball is above a height of 2 m.

Step 1: Draw a clear diagram to represent the information given.

Defining up as positive:

$+$

$u = 15$

$a = -9.8$

Step 2: Fill the information in *suvat*, identifying what is required with a question mark.

a $s = 0$
$u = 15$
v not required
$a = -9.8$
$t = ?$

b $s = 2$
$u = 15$
v not required
$a = -9.8$
$t = ?$

Step 3: Pick an equation of motion relating the three known variables with the unknown that is required, insert values, rearrange (if necessary) and solve.

$s = ut + \frac{1}{2}at^2$
$0 = 15t - 4.9t^2$
$0 = t(15 - 4.9t)$
$t = 0$ or $3.06\ldots$

The time of flight is
3.1 s (2 s.f.).

$s = ut + \frac{1}{2}at^2$
$2 = 15t - 4.9t^2$
$0 = 4.9t^2 - 15t + 2$
$t = 0.140\ldots$ or $2.92\ldots$

The ball is above 2 m for
$2.9 - 0.14 = 2.8$ s (2 s.f.).

Tip:
These are the two times for which the particle is at the given displacement.

Example 2.14 A man throws a ball vertically upwards with a speed of 15 m s^{-1} from a height of 1.5 m above the ground. The ball travels freely under gravity.

a Calculate the time of flight.

b Calculate the speed with which the ball hits the ground.

Step 1: Draw a clear diagram to represent the information given.

Defining up as positive:

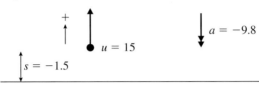

Step 2: Fill the information in *suvat*, identifying what is required with a question mark.

a
$s = -1.5$
$u = 15$
v not required
$a = -9.8$
$t = ?$

b
$s = -1.5$
$u = 15$
$v = ?$
$a = -9.8$
t not required

Note:
$s = 0$ at the height from which the ball is thrown and $s = -1.5$ at the ground.

Step 3: Pick an equation of motion relating the three known variables with the unknown that is required, insert values, rearrange (if necessary) and solve.

a
$s = ut + \frac{1}{2}at^2$
$-1.5 = 15t - 4.9t^2$
$4.9t^2 - 15t - 1.5 = 0$
$t = 3.15\ldots$ or $-0.096\ldots$

The time of flight is 3.2 s (2 s.f.).

b
$v^2 = u^2 + 2as$
$v^2 = 15^2 + 2(-9.8)(-1.5)$
$v = -15.9$

The speed with which the ball hits the ground is 16 m s^{-1} (2 s.f.).

Note:
The similarity between Example 2.13a and Example 2.14a.

SKILLS CHECK **2B: Motion in a vertical plane**

1 A ball is thrown vertically upwards with velocity 28 m s^{-1} and travels freely under gravity. Find the velocity after two seconds and the distance that the particle has travelled from the start at this time.

2 Find the maximum height reached and the time of flight (the time taken to reach the start again) for the ball in question **1**. For how long is the ball above a height of 20 m?

3 A book is dropped from 80 m above ground and travels freely under gravity. Find the time taken for it to reach the ground. What assumptions have you made when modelling this situation?

 4 A stone is thrown vertically upwards from 4 m above ground. The initial velocity of the stone is 14 m s^{-1}. Find:

a the maximum height above the ground reached by the stone

b the time taken for it to hit the ground.

5 A ball is thrown upwards and reaches a height of 50 m. Find its initial velocity.

 6 A ball is dropped from a window at $t = 0$, where t is the time in seconds. At $t = 2$ another ball is thrown from the same point with downwards velocity 25 m s^{-1}. Given that the balls travel freely under gravity, find:

a the time when the balls pass each other

b the distance from the window when they pass each other.

7 a Show that $h = \dfrac{u^2}{2g}$ for a particle travelling freely under gravity in the vertical plane with an initial velocity of u m s^{-1} (upwards) and a maximum height of h m above its starting point, where g is the acceleration due to gravity.

b Calculate the difference in the maximum heights of two particles, travelling freely under gravity, given they are thrown upwards with velocities 10 and 15 m s^{-1} respectively.

SKILLS CHECK **2B EXTRA is on the CD**

2.6 Vectors

Magnitude and direction of quantities represented by a vector.

A **vector quantity** is one that has both size and direction, as opposed to a **scalar quantity** that has size only. When you add two scalar quantities, you can add the numbers directly. When you add two vector quantities, you must take their directions into account first.

Example 2.15 A man walks 2 km due north from point O to point A and then 3 km due east to point B. Find the distance OB and the bearing of the point B from O.

Step 1: Draw a clear diagram, marking all known angles and distances.

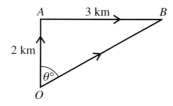

Note:
Bearings are measured clockwise from north (in this case the line OA).

$$OB^2 = OA^2 + AB^2$$
$$= 2^2 + 3^2 = 13$$
$$OB = \sqrt{13} \text{ km}$$

Tip:
OAB is a right-angled triangle. Solve using Pythagoras' theorem.

Step 2: Use trigonometry to calculate unknown angles and distances.

$$\tan \theta° = 3/2$$
$$\theta = 56.3\ldots$$

The bearing is $056°$ (nearest degree).

Example 2.16 The point A is on a bearing of $045°$ and at a distance of 25 m from the point O. The point B is on a bearing of $160°$ and at a distance of 40 m from the point A. Find the distance OB and the bearing of B from O.

Step 1: Draw a clear diagram, marking all known angles and distances.

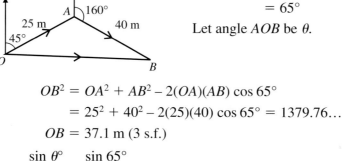

Angle $OAB = 360° - 135° - 160°$
$$= 65°$$
Let angle AOB be θ.

Recall:
Interior angles on parallel lines add up to $180°$ and angles around a point add up to $360°$.

Step 2: Use trigonometry to calculate unknown angles and distances.

$$OB^2 = OA^2 + AB^2 - 2(OA)(AB) \cos 65°$$
$$= 25^2 + 40^2 - 2(25)(40) \cos 65° = 1379.76\ldots$$
$$OB = 37.1 \text{ m (3 s.f.)}$$

$$\frac{\sin \theta°}{AB} = \frac{\sin 65°}{OB}$$

$$\sin \theta° = \frac{40 \sin 65°}{37.1\ldots} = 0.9759\ldots$$

$$\theta = 77.4\ldots$$

Tip:
OAB is not a right-angled triangle. Find OB using the cosine rule.
Find θ using the sine rule.
The sine and cosine rules are covered in C2.

The bearing of B from O is $(45° + 77.4°) = 122°$ (nearest degree).

Notation: In the above example the **distance** OA is 25 m, while the **vector** OA is 25 m on a bearing of $045°$. To distinguish between the distance OA and the vector OA an arrow is placed over the letters to represent the vector, thus: $\overrightarrow{OA}$.

Note:
The direction of the arrow over the letters describes the direction of the vector.

Example 2.16 can be described in this notation as:

$$\overrightarrow{OB} = \overrightarrow{OA} + \overrightarrow{AB}$$

The vector $\overrightarrow{OB}$ is called the **resultant** of the vectors $\overrightarrow{OA}$ and $\overrightarrow{AB}$.

Notation: You can also represent vectors using a single letter in bold type. If the vector $\overrightarrow{OA} = \mathbf{x}$, the vector $\overrightarrow{AB} = \mathbf{y}$, and the vector $\overrightarrow{OB} = \mathbf{z}$, then:

$$\mathbf{z} = \mathbf{x} + \mathbf{y}$$

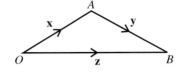

Note:
If the vector from O to $A = \mathbf{x}$, then the vector from A to $O = -\mathbf{x}$
i.e. $\overrightarrow{OA} = \mathbf{x}$
$\overrightarrow{AO} = -\mathbf{x}$

The size of the vector $\mathbf{x}$ can be written as $|\mathbf{x}|$ or simply x without the bold type. This refers to the **scalar magnitude**: the distance OA.

If two vectors, $\mathbf{x}$ and $\mathbf{y}$, are equal then you can write:

$$\mathbf{x} = \mathbf{y}$$

Note:
This means that they both have the same size and direction.

If two vectors, $\mathbf{x}$ and $\mathbf{y}$, are parallel but have different magnitudes, then you can write:

$$\mathbf{x} = k\mathbf{y}$$

where k is a scalar.

Note:
This means that they both have the same direction, but not necessarily the same size: the vector $\mathbf{x}$ is k times bigger than the vector $\mathbf{y}$.

When you handwrite vectors, since you cannot show bold type, you underline the letter, for example <u>a</u> or <u>a</u>.

2.7 The unit vectors, i and j

Use of unit vectors **i** and **j**.

The i and j vectors

Vector $\mathbf{i}$ represents one unit in the positive direction along the x-axis, vector $\mathbf{j}$ represents one unit in the positive direction along the y-axis, where x and y are perpendicular.

For example, the vector $2\mathbf{i} - 3\mathbf{j}$ from the origin can be drawn as:

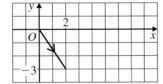

Note:
The $-3\mathbf{j}$ means that you move 3 units in the negative direction along the y-axis.

Sometimes the vector $a\mathbf{i} + b\mathbf{j}$ can be represented as $\begin{bmatrix} a \\ b \end{bmatrix}$. This is called a **column vector**.

So $2\mathbf{i} - 3\mathbf{j} \equiv \begin{bmatrix} 2 \\ -3 \end{bmatrix}$.

The size and direction of the vector $a\mathbf{i} + b\mathbf{j}$

If the vector $\mathbf{v} = a\mathbf{i} + b\mathbf{j}$, then you can find the size of $\mathbf{v}$, written $|\mathbf{v}|$ or v, by using Pythagoras' theorem.

$$|\mathbf{v}| = \sqrt{a^2 + b^2}$$

and $\tan \theta° = \dfrac{b}{a}$

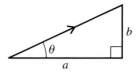

where $\theta°$ is the angle that the vector makes with the positive x-axis in an anticlockwise direction. Use a diagram to find the acute angle made with the x-axis and then find the angle made with the positive x-axis.

Note:
The size of a vector can also be called the **magnitude** or **modulus**.

Example 2.17 Calculate the modulus, $|\mathbf{v}|$, and the angle anticlockwise from the positive x-axis when:

a $\mathbf{v} = 2\mathbf{i} + 4\mathbf{j}$ **b** $\mathbf{v} = 3\mathbf{i} - \mathbf{j}$

Step 1: Draw a clear diagram, marking all known angles and distances.

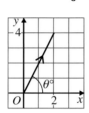   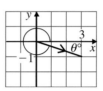

Step 2: Use trigonometry to calculate unknown angles and distances.

$|\mathbf{v}| = \sqrt{2^2 + 4^2} = 4.47$ (3 s.f.) $\quad$ $|\mathbf{v}| = \sqrt{3^2 + (-1)^2} = 3.16$ (3 s.f.)

$\tan \theta° = \dfrac{4}{2}$ $\qquad\qquad\qquad\qquad$ $\tan \theta° = \dfrac{1}{3}$

$\quad \theta = 63.4\ldots$ $\qquad\qquad\qquad\qquad$ $\theta = 18.4\ldots$

The angle between the vector and the positive x-axis is 63.4° (3 s.f.)

The angle between the vector and the positive x-axis is $360° - 18.4\ldots° = 342°$ (3 s.f.)

Recall:
$|\mathbf{v}| = \sqrt{a^2 + b^2}$

Note:
Imagine the vector makes a right-angled triangle with the axis, solve for the acute angle, then calculate the angle anti-clockwise from the x-axis.

The i and j components of any vector

A vector of modulus $|\mathbf{M}|$, or M, that acts at an angle θ to the positive x-axis can also be expressed in the form $x\mathbf{i} + y\mathbf{j}$. You can calculate these horizontal and vertical **components** using simple trigonometry:

The horizontal component, $x\mathbf{i}$:

$\cos \theta° = \dfrac{x}{M}$

$x = M \cos \theta°$

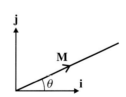

The vertical component, $y\mathbf{j}$:

$\sin \theta° = \dfrac{y}{M}$

$y = M \sin \theta$

$$x\mathbf{i} + y\mathbf{j} = M \cos \theta°\mathbf{i} + M \sin \theta°\mathbf{j}$$

Example 2.18 **a** The vector OA has a magnitude of 4 units and acts at an angle of 30° to the vector $\mathbf{i}$. Find the vector OA in the form $x\mathbf{i} + y\mathbf{j}$.

b The vector OB has magnitude 6 units and acts at an angle of 130° to the vector $\mathbf{i}$. Find the vector OB in the form $x\mathbf{i} + y\mathbf{j}$.

Note:
Let O be at the origin.

Step 1: Draw a clear diagram, marking all known angles and distances.

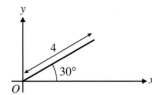

a

$$x\mathbf{i} + y\mathbf{j} = M\cos\theta\mathbf{i} + M\sin\theta\mathbf{j}$$
$$= 4\cos30°\mathbf{i} + 4\sin30°\mathbf{j}$$
$$= 2\sqrt{3}\mathbf{i} + 2\mathbf{j}$$

Step 2: Use trigonometry to calculate unknown angles and distances.

b

$$x\mathbf{i} + y\mathbf{j} = -6\cos(50°)\mathbf{i} + 6\sin(50°)\mathbf{j}$$
$$= -3.86\mathbf{i} + 4.60\mathbf{j} \text{ (3 s.f.)}$$

Tip:
First calculate the acute angle made with the axis then indicate the direction as +ve or −ve.

Adding and subtracting vectors

$$(a_1\mathbf{i} + a_2\mathbf{j}) + (b_1\mathbf{i} + b_2\mathbf{j}) = (a_1 + b_1)\mathbf{i} + (a_2 + b_2)\mathbf{j}$$

You add the **i** components and the **j** components separately.

$$(a_1\mathbf{i} + a_2\mathbf{j}) - (b_1\mathbf{i} + b_2\mathbf{j}) = (a_1 - b_1)\mathbf{i} + (a_2 - b_2)\mathbf{j}$$

You subtract the **i** components and the **j** components separately.

Note:
In column vector form, this could be written

$$\begin{bmatrix} a_1 \\ a_2 \end{bmatrix} + \begin{bmatrix} b_1 \\ b_2 \end{bmatrix} = \begin{bmatrix} a_1 + b_1 \\ a_2 + b_2 \end{bmatrix}$$

SKILLS CHECK **2C: Vectors**

1 A man walks x km from O on a bearing of $\theta°$ to a point A. He then walks a further distance of y km on a bearing of $\alpha°$ to a point B. Find the distance OB and the bearing of O from B when:

  a $\theta = 090$, $\alpha = 180$, $x = 4$ and $y = 3$

  b $\theta = 024$, $\alpha = 160$, $x = 7$ and $y = 2$

  c $\theta = 290$, $\alpha = 045$, $x = 5$ and $y = 2$

 2 Find the modulus and the angle that each of the following vectors makes with the vector **i**:

  a $3\mathbf{i} + 4\mathbf{j}$ **b** $6\mathbf{i} - \mathbf{j}$ **c** $\begin{bmatrix} -2 \\ -2 \end{bmatrix}$ **d** $\begin{bmatrix} -3 \\ 6 \end{bmatrix}$

3 Find the following vectors in the form $x\mathbf{i} + y\mathbf{j}$ which have magnitude M and act at an angle $\theta°$ to the vector **i** where:

  a $M = 9$, $\theta = 20$ **b** $M = 4$, $\theta = 170$ **c** $M = 14$, $\theta = 210$

4 Given the vector $\mathbf{a} = 3\mathbf{i} + \mathbf{j}$, the vector $\mathbf{b} = 2\mathbf{i} - 4\mathbf{j}$ and the vector $\mathbf{c} = x\mathbf{i} + y\mathbf{j}$, find **c** in terms of **i** and **j**, the magnitude of **c** and the smaller angle that **c** makes with the vector **j** if:

  a $\mathbf{c} = \mathbf{a} + \mathbf{b}$ **b** $\mathbf{c} = \mathbf{a} - \mathbf{b}$ **c** $\mathbf{c} = 2\mathbf{a} + \mathbf{b}$

  d $\mathbf{c} = 2\mathbf{b} - 3\mathbf{a}$ **e** $\mathbf{c} + \mathbf{b} = 2\mathbf{a}$

SKILLS CHECK **2C EXTRA** is on the CD

2.8 Application of vectors

Application of vectors in two dimensions to represent position, velocity or acceleration.

The displacement vector

The vectors used in Section 2.6 are **displacement vectors**; they describe relative positions. When measured from the origin the displacement vector is called a **position vector**.

Example 2.19   A particle is at point A (3, 4) in the $x-y$ plane. After two seconds the particle has moved to point B, where $\overrightarrow{AB} = 3\mathbf{i} - \mathbf{j}$. Find the position vector $\overrightarrow{OB}$.

Step 1: Use the vector formula and simplify.

$$\overrightarrow{OB} = \overrightarrow{OA} + \overrightarrow{AB}$$
$$= 3\mathbf{i} + 4\mathbf{j} + 3\mathbf{i} - \mathbf{j} = (6\mathbf{i} + 3\mathbf{j})$$

Note:
The coordinates (3, 4) represent the position vector $(3\mathbf{i} + 4\mathbf{j})$.

The velocity vector

Velocity is a vector quantity; it is the rate of change of displacement with time. The magnitude of velocity is called the **speed**.
When the velocity is constant (no acceleration):

$$\text{velocity} = \frac{\text{change in displacement}}{\text{time taken}}$$

A particle moving $(x\mathbf{i} + y\mathbf{j})$ m every second has a **velocity** $(x\mathbf{i} + y\mathbf{j})$ m/s.

Note:
This is the vector form of the scalar relationship:
$$\text{speed} = \frac{\text{distance}}{\text{time}}$$

Note:
You can also write m/s as m s^{-1}.

Example 2.20   At time $t = 0$ (in seconds), a particle is at a position $\begin{bmatrix} 6 \\ -7 \end{bmatrix}$ m relative to the origin O. At $t = 3$ the particle is at a position $\begin{bmatrix} 3 \\ 2 \end{bmatrix}$ m relative to the origin O. Given that the velocity is constant find:

a   the velocity   **b**   the speed of the particle.

Step 1: Use the vector formula.

a $\mathbf{v} = \dfrac{\begin{bmatrix} 3 \\ 2 \end{bmatrix} - \begin{bmatrix} 6 \\ -7 \end{bmatrix}}{3}$

Step 2: Simplify.

$$= \frac{1}{3}\begin{bmatrix} -3 \\ 9 \end{bmatrix}$$

$$= \begin{bmatrix} -1 \\ 3 \end{bmatrix}$$

The velocity is $\begin{bmatrix} -1 \\ 3 \end{bmatrix}$ m s^{-1}.

b $|\mathbf{v}| = \left\| \begin{bmatrix} -1 \\ 3 \end{bmatrix} \right\| = \sqrt{(-1)^2 + 3^2}$

$$= 3.16 \text{ (3 s.f.)}$$

The speed is 3.16 m s^{-1}.

Tip:
Don't forget the units.

Tip:
It is often easier to use column vectors. You can be examined on both the $\mathbf{i}-\mathbf{j}$ form and the column vector form.

The acceleration vector

The acceleration of a particle is the rate of change of the velocity with time. When the acceleration is constant:

$$\text{acceleration} = \frac{\text{change in velocity}}{\text{time taken}}$$

The units for acceleration are m/s^2 or m s^{-2}.

Example 2.21   A particle is initially travelling with velocity $(-2\mathbf{i} - 9\mathbf{j})$ m s^{-1} and 2 seconds later it has velocity $(6\mathbf{i} - 11\mathbf{j})$ m s^{-1}, where $\mathbf{i}$ and $\mathbf{j}$ are unit vectors in the directions of the positive x- and y-axes, respectively. Given that the acceleration of the particle is constant, find:

a   the acceleration   **b**   the magnitude of the acceleration
c   the angle that the acceleration makes with the vector $\mathbf{j}$.

Step 1: Use the vector formula.

a $\mathbf{a} = \dfrac{(6\mathbf{i} - 11\mathbf{j}) - (-2\mathbf{i} - 9\mathbf{j})}{2}$

Step 2: Simplify.

$$= \frac{8\mathbf{i} - 2\mathbf{j}}{2}$$

$$= (4\mathbf{i} - \mathbf{j})$$

The acceleration is $(4\mathbf{i} - \mathbf{j})$ m s^{-2}.

b $|\mathbf{a}| = |4\mathbf{i} - \mathbf{j}| = \sqrt{4^2 + (-1)^2}$

$$= \sqrt{17}$$

The magnitude of the acceleration is $\sqrt{17}$ m s^{-2}.

Step 1: Draw a clear diagram, marking all known angles and distances.

Step 2: Use trigonometry to calculate unknown angles.

c $\tan \theta° = \frac{1}{4}$

$\quad \theta = 14$ (nearest degree)

The angle made with the vector **j** is $(90° + 14°) = 104°$.

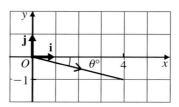

2.9 Equations of motion with vectors

Finding position, velocity, speed and acceleration of a particle moving in two dimensions with constant acceleration.

The equations of motion can also be applied to vectors, but the displacement, velocity and acceleration are given as vectors. The equations of motion (using the variables **r**, **u**, **v**, **a**, t) are given below:

$\mathbf{r} = \mathbf{u}t + \frac{1}{2}\mathbf{a}t^2$

$\mathbf{r} = \mathbf{v}t - \frac{1}{2}\mathbf{a}t^2$

$\mathbf{v} = \mathbf{u} + \mathbf{a}t$

$\mathbf{r} = \frac{1}{2}(\mathbf{u} + \mathbf{v})t$

where **r** is the displacement vector, **u** is the initial velocity vector, **v** is the final velocity vector, and **a** is the acceleration vector. Many of the properties of vectors are applied in these questions.

> **Note:**
> These are the same as the *suvat* equations on page 8 with the exceptions that
> i s is replaced with **r**
> ii the equation $\mathbf{v}^2 = \mathbf{u}^2 + 2\mathbf{a}s$ is omitted because there is no definition of squaring a vector.

Example 2.22 A particle leaves a point O and travels for 4 seconds with constant acceleration $(2\mathbf{i} - 2\mathbf{j})$ m s^{-2}, where **i** and **j** are unit vectors due east and due north respectively.

    a Given that the final velocity of the particle is $(9\mathbf{i} - 5\mathbf{j})$ m s^{-1}, find the displacement of the particle from O after 4 seconds.

    b Hence, find the distance that the particle is from O after 4 seconds.

> **Note:**
> You should learn these formulae.

Step 1: Fill the information in **ruvat**, identifying what is required with a question mark.

Using column vector notation:

$\mathbf{r} = ?$

$\mathbf{u} =$

$\mathbf{v} = \begin{bmatrix} 9 \\ -5 \end{bmatrix}$

$\mathbf{a} = \begin{bmatrix} 2 \\ -2 \end{bmatrix}$

$t = 4$

> **Note:**
> These steps are the same as before except there is no need to draw a diagram.

Step 2: Pick an equation of motion relating the three knowns with the unknown that is required. Insert values and solve.

a $\mathbf{r} = \mathbf{v}t - \frac{1}{2}\mathbf{a}t^2$

$\mathbf{r} = 4\begin{bmatrix} 9 \\ -5 \end{bmatrix} - \frac{1}{2}\begin{bmatrix} 2 \\ -2 \end{bmatrix}4^2$

$\quad = \begin{bmatrix} 20 \\ -4 \end{bmatrix}$

The displacement of the particle from O after 4 seconds is $(20\mathbf{i} - 4\mathbf{j})$ m.

b The distance of the particle from O after 4 seconds is $\sqrt{20^2 + (-4)^2} = 20.4$ m (3 s.f.).

> **Recall:**
> The distance is the magnitude of the displacement.

Example 2.23 A particle is initially travelling with velocity $(-2\mathbf{i} - 9\mathbf{j})$ m s^{-1} and 2 seconds later it has velocity $(6\mathbf{i} - 11\mathbf{j})$ m s^{-1}, where $\mathbf{i}$ and $\mathbf{j}$ are unit vectors in the directions of the positive x- and y-axes respectively.

   a Find the acceleration of the particle, given that it is constant.

   b Given that the particle is at the origin at $t = 0$ seconds, find the position vector of the particle after 2 seconds. Hence find the distance of the particle from the origin after 2 seconds.

Step 1: Fill the information in **ruvat**, identifying what is required with a question mark.

$$\mathbf{r} = ?$$
$$\mathbf{u} = (-2\mathbf{i} - 9\mathbf{j})$$
$$\mathbf{v} = (6\mathbf{i} - 11\mathbf{j})$$
$$\mathbf{a} = ?$$
$$t = 2$$

Step 2: Pick an equation of motion relating the three knowns with the unknown that is required. Insert values and solve.

a
$$\mathbf{v} = \mathbf{u} + \mathbf{a}t$$
$$6\mathbf{i} - 11\mathbf{j} = -2\mathbf{i} - 9\mathbf{j} + 2\mathbf{a}$$
$$2\mathbf{a} = 6\mathbf{i} - 11\mathbf{j} + 2\mathbf{i} + 9\mathbf{j}$$
$$= 8\mathbf{i} - 2\mathbf{i}$$
$$\mathbf{a} = 4\mathbf{i} - \mathbf{j}$$

The acceleration of the particle is $(4\mathbf{i} - \mathbf{j})$ m s^{-2}.

> **Note:**
> You could have used
> $\mathbf{r} = \frac{1}{2}(\mathbf{u} + \mathbf{v})t$.

b
$$\mathbf{r} = \mathbf{u}t + \frac{1}{2}\mathbf{a}t^2$$
$$= (-2\mathbf{i} - 9\mathbf{j})(2) + \frac{1}{2}(4\mathbf{i} - \mathbf{j})(2^2)$$
$$= -4\mathbf{i} - 18\mathbf{j} + 8\mathbf{i} - 2\mathbf{j}$$
$$= 4\mathbf{i} - 20\mathbf{j}$$

The position vector of the particle at $t = 2$ is $(4\mathbf{i} - 20\mathbf{j})$ m.
The distance from the origin at $t = 2$ is $\sqrt{(4^2 + 20^2)} = 20.4$ m.

Example 2.24 A car moves with constant acceleration so that its velocity, in m s^{-1}, at time t seconds, is given by

$$\mathbf{v} = (2t - 5)\mathbf{i} + (4 - t)\mathbf{j}$$

where $\mathbf{i}$ and $\mathbf{j}$ are unit vectors directed due east and due north respectively.

   a By comparing the velocity vector of the car with $\mathbf{v} = \mathbf{u} + \mathbf{a}t$, find the acceleration of the car.

   b Find the times at which the car is moving:

     i due east

     ii north-east.

Step 1: Rewrite in the form $\mathbf{v} = \mathbf{u} + \mathbf{a}t$, and identify the acceleration.

a
$$\mathbf{v} = (2t - 5)\mathbf{i} + (4 - t)\mathbf{j}$$
$$= (-5\mathbf{i} + 4\mathbf{j}) + (2\mathbf{i} - \mathbf{j})t$$

So the acceleration of the car is $(2\mathbf{i} - \mathbf{j})$ m s^{-2}.

> **Note:**
> The initial velocity of the particle is $(-5\mathbf{i} + 4\mathbf{j})$ m s^{-1}.

Step 2: Use the properties of vectors and solve for time.

b **i** Setting the $\mathbf{j}$ component equal to 0:
$$4 - t = 0$$
$$t = 4$$

The particle is moving due east at time $t = 4$.

> **Recall:**
> If a particle is moving due east, then the vector has no $\mathbf{j}$ component at this time.

   ii Equating the $\mathbf{i}$ and $\mathbf{j}$ components:
$$2t - 5 = 4 - t$$
$$t = 3$$

The particle is moving north-east at time $t = 3$.

> **Recall:**
> The particle moves north-east when the $\mathbf{i}$ and $\mathbf{j}$ components are equal.

Problems involving resultant velocities.

Resultant velocity

When velocities are given in vector form the **resultant velocity** can be found by adding the velocities together, using vector addition.

Example 2.25

Find the resultant velocity of the velocities $(2\mathbf{i} - \mathbf{j})$ m s^{-1} and $(3\mathbf{i} + 4\mathbf{j})$ m s^{-1}. Hence, find the magnitude of the resultant and the angle that it makes with the vector $\mathbf{i}$.

Step 1: Find the resultant for each component separately; add the $\mathbf{i}$ parts and the $\mathbf{j}$ parts together.

Resultant velocity,

$$\mathbf{v} = (2\mathbf{i} - \mathbf{j}) + (3\mathbf{i} + 4\mathbf{j})$$
$$= (5\mathbf{i} + 3\mathbf{j})$$

The resultant velocity is $(5\mathbf{i} + 3\mathbf{j})$ m s^{-1}.

Step 2: Find the overall resultant of the two (vector) components using trigonometry.

$$|\mathbf{v}| = \sqrt{5^2 + 3^2} = 5.83$$

$$\tan \theta° = \frac{3}{5}$$

$$\theta = 31.0$$

The magnitude of the resultant velocity is 5.83 m s^{-1} and it makes an angle of 31.0° with the vector $\mathbf{i}$.

> **Recall:**
> If $\mathbf{v} = a\mathbf{i} + b\mathbf{j}$ then
> $|\mathbf{v}| = \sqrt{a^2 + b^2}$ and $\tan \theta° = \frac{b}{a}$.

Velocity triangles

When a particle travels with a given velocity through the air, where there is a wind present, or through water, where there is a current, then you can find the resultant velocity using vector triangles, where the resultant velocity is found using simple trigonometry:

> **Note:**
> If the velocities are given in vector form then you can calculate the resultant velocity as in the previous example.

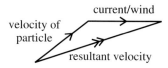

Example 2.26

A boat travels across a river from one bank to the other. The downstream current has speed 4 m s^{-1} and the speed of the boat in still water is 6 m s^{-1}.

a If the boat sets course to cross the river perpendicular to the bank, find the resultant velocity V m s^{-1} and the angle $\theta°$ it makes with the bank.

b If the boat sets course at an angle of $\theta°$ to the bank, so that its resultant velocity will be perpendicular to the bank, then find the resultant velocity and the angle $\theta°$.

Step 1: Draw a velocity triangle with the resultant marked with a double arrow.

a

Step 2: Find the resultant and angle using trigonometry.

Using Pythagoras' theorem, $V = \sqrt{(6^2 + 4^2)} = 7.21$ (3 s.f.).

If $\theta°$ is the angle that the resultant velocity makes with the bank then $\tan \theta° = \frac{6}{4}$ and $\theta = 56.3$. So the resultant velocity is 7.21 m s^{-1} and the angle it makes with the bank is 56.3°.

Step 1: Draw a velocity triangle with the resultant marked with a double arrow.

b You require the resultant velocity to be perpendicular to the bank.

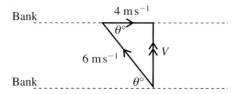

Step 2: Find the resultant and angle using trigonometry.

Using Pythagoras' theorem, $V = \sqrt{(6^2 - 4^2)} = 4.47$ (3 s.f.).

If $\theta°$ is the angle that the boat's direction makes with the bank then $\cos\theta° = \frac{4}{6}$ and $\theta = 48.2$.

So the resultant velocity is 4.47 m s^{-1} and the boat should set off at an angle of $48.2°$ to the bank.

Example 2.27 A plane wishes to set a course from P to Q, where Q is 800 km due east of P. There is a wind of 60 km h^{-1} blowing from the north-west. The plane has speed 500 km h^{-1} in still air.

a Find the course that the plane must set.

b By finding the resultant velocity, find the time taken for the plane to travel from P to Q.

Step 1: Draw a velocity triangle with the resultant marked with a double arrow.

Let V = resultant velocity

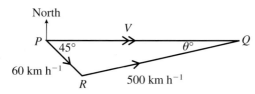

Step 2: Find the resultant and angle using trigonometry.

a In triangle PQR, using the sine rule:

$$\frac{\sin\theta°}{60} = \frac{\sin 45°}{500}$$

$$\sin\theta° = 0.0848\ldots$$

$$\theta = 4.87\ldots = 5° \text{ (nearest degree)}$$

So angle $PRQ = 180° - (45° + 5°) = 130°$ and so the bearing of Q from $R = 130° - 45° = 85°$.

So the course that the plane must set is $085°$.

b Using the sine rule again, you can find V:

$$\frac{V}{\sin 130°} = \frac{500}{\sin 45°}$$

$$V = 541$$

Step 3: Use speed $= \dfrac{\text{distance}}{\text{time}}$.

Rearrange the formula speed $= \dfrac{\text{distance}}{\text{time}}$ to find the time taken.

$$\text{Time} = \frac{\text{distance}}{\text{speed}}$$

$$= \frac{800}{540.6\ldots}$$

$$= 1.48 \text{ (3 s.f.)}$$

The time taken is 1.48 hours.

1 Find the constant velocity and the speed of a particle that moves from position vectors $\mathbf{r}_A$ m to $\mathbf{r}_B$ m in t seconds where:

  a $\mathbf{r}_A = 5\mathbf{i} + \mathbf{j}$, $\mathbf{r}_B = 7\mathbf{i} + 3\mathbf{j}$, $t = 2$

  b $\mathbf{r}_A = 5\mathbf{i} - \mathbf{j}$, $\mathbf{r}_B = 2\mathbf{i} + 5\mathbf{j}$, $t = 3$

  c $\mathbf{r}_A = 4\mathbf{i} - 6\mathbf{j}$, $\mathbf{r}_B = 13\mathbf{i} + 24\mathbf{j}$, $t = 6$.

2 A particle travels with constant velocity $(7\mathbf{i} + 8\mathbf{j})$ m s^{-1} starting from the point with position vector $(3\mathbf{i} + \mathbf{j})$ m relative to a fixed origin O. Find:

  a the displacement of the particle after 3 seconds

  b the position vector of the particle after 3 seconds

  c the distance that the particle is from O after 3 seconds and the angle that the position vector makes with the vector $\mathbf{i}$.

3 A car travels with uniform acceleration. Initially the car has velocity $\mathbf{v}_1$ m s^{-1} and after t seconds it has velocity $\mathbf{v}_2$ m s^{-1}. Find the acceleration vector of the particle where:

  a $\mathbf{v}_1 = (6\mathbf{i} + 3\mathbf{j})$, $\mathbf{v}_2 = (9\mathbf{i} + 12\mathbf{j})$, $t = 2$

  b $\mathbf{v}_1 = (-77\mathbf{i} + 32\mathbf{j})$, $\mathbf{v}_2 = (-13\mathbf{i} - 32\mathbf{j})$, $t = 4$

  c $\mathbf{v}_1 = (-\mathbf{i} + \mathbf{j})$, $\mathbf{v}_2 = (-\mathbf{i} + 6\mathbf{j})$, $t = 5$.

 4 A ball is rolling across a smooth table. It accelerates uniformly at $(2\mathbf{i} + \mathbf{j})$ m s^{-2}, where $\mathbf{i}$ and $\mathbf{j}$ are unit vectors due east and due north respectively. The final velocity of the ball after travelling with this acceleration for 4 seconds is $(-4\mathbf{i} - 12\mathbf{j})$ m s^{-1}. Find the initial speed of the ball.

5 At $t = 0$, where t is the time measured in seconds, a particle sets out from a point A with position vector $(3\mathbf{i} + 8\mathbf{j})$ m and travels with velocity $(\mathbf{i} - 2\mathbf{j})$ m s^{-1}.

  a Find the position vector of the particle after t seconds.

  After three seconds the particle reaches the point B.

  b Find the position vector of the point B.

  c Show that the distance of OB, where O is the origin, can be written as $a\sqrt{10}$ m. State the value of a.

6 A car moves with constant acceleration. Initially, the car is travelling due east at 2 m s^{-1} and 2 seconds later it is travelling due north at 4 m s^{-1}. The unit vectors $\mathbf{i}$ and $\mathbf{j}$ are directed due east and north respectively.

  a Write down the initial velocity and the final velocity of the car after it has travelled for 2 seconds.

  b What is the magnitude of the acceleration of the car?

  c Assuming that the car starts from the origin, find its position after 2 seconds.

 7 (In this question $\mathbf{i}$ and $\mathbf{j}$ are perpendicular unit vectors along a horizontal plane.) Initially, a particle travelling along a smooth horizontal plane is at the origin and travels with velocity $(2\mathbf{i} + 5\mathbf{j})$ m s^{-1}. After 8 seconds it travels with velocity $(10\mathbf{i} - 3\mathbf{j})$ m s^{-1}. Given that the acceleration of the particle is constant:

  a find the acceleration of the particle

  b find the position vector of the particle when its velocity is $(12\mathbf{i} - 5\mathbf{j})$ m s^{-1}

  c hence, find the distance that the particle is from the origin when its velocity is $(12\mathbf{i} - 5\mathbf{j})$ m s^{-1}.

8 A particle moves so that its position vector, in metres, relative to an origin O after t seconds is given by

$$\mathbf{r} = (4t - 7)\mathbf{i} + (3 - t)\mathbf{j}$$

where $\mathbf{i}$ and $\mathbf{j}$ are unit vectors directed due east and due north respectively.

a Find the distance of the particle from O at $t = 10$.

b Find the times when the particle is

  i due east of O **ii** north-east of O.

9 a Find the resultant velocity of the following velocities: $(\mathbf{i} - \mathbf{j})$ and $(2\mathbf{i} - 7\mathbf{j})$ m s^{-1}. Hence, find the magnitude of the resultant and the angle that it makes with the vector $\mathbf{j}$.

b If the resultant velocity of the velocities $(14\mathbf{i} - 11\mathbf{j})$ and $(x\mathbf{i} + y\mathbf{j})$ m s^{-1} is $(12\mathbf{i} - 3\mathbf{j})$ m s^{-1}, find x and y.

10 A boat travels across a river from one bank to the other, where the banks are parallel to each other. The downstream current has speed 5 m s^{-1} and the speed of the boat in still water is 12 m s^{-1}.

a If the boat sets course to directly cross the river at right angles to the bank, find the resultant velocity and the angle it makes with the bank.

b If the boat wishes to reach the opposite bank directly ahead, then the resultant velocity must be perpendicular to the bank. Find the resultant velocity and the angle that the boat's direction should make with the bank from which it leaves.

11 A plane sets a course due east. However, there is a wind that blows due south at 10 m s^{-1}. Given that the plane can fly at 140 km h^{-1} when there is no wind, find the resultant velocity of the plane and its bearing.

SKILLS CHECK **2D EXTRA is on the CD**

Examination practice Kinematics in one and two dimensions

1 A car is initially at rest on a straight, horizontal road. It accelerates uniformly for 8 seconds, reaching a speed of 20 m s^{-1}. It then travels at this constant speed for a further 40 seconds.

a Sketch a velocity–time graph to illustrate the motion of the car.

b Find the total distance travelled by the car in the 48 seconds.

c Show that the acceleration of the car during the first 8 seconds is 2.5 m s^{-2}.   [AQA (B) Nov 2000]

2

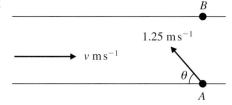

A girl swims across a river. When she swims in still water, she swims at 1.25 m s^{-1}. The river flows parallel to its banks at v m s^{-1}.

The girl aims to swim upstream at an angle θ degrees to the river bank so that her resultant velocity, of magnitude 1 m s^{-1}, is along AB, perpendicular to the river bank.

a Sketch an appropriate triangle of velocities.

b Find the value of v.

c Find the value of θ.   [AQA (A) Jan 2003]

3 At time $t = 0$, a boat is travelling due east at a speed of 3 m s^{-1}. The unit vectors **i** and **j** are directed east and north respectively.

    a Write down the initial velocity of the boat in vector form.

    b The boat has a constant acceleration of $(0.1\mathbf{i} + 0.2\mathbf{j})$ m s^{-2}. Find an expression for the velocity of the boat at time t seconds.

    c When $t = T$, the boat is travelling north east. Form an equation that T must satisfy, and solve it to show that $T = 30$.

    d Find the distance of the boat from its initial position when $t = 20$.   [AQA (B) Nov 2002]

4 A particle moves in the horizontal plane that contains the perpendicular unit vectors **i** and **j**. Initially it is at the origin and has velocity 18**i** m s^{-1}. After accelerating for 10 seconds its velocity is $(30\mathbf{i} + 8\mathbf{j})$ m s^{-1}. Assume that the acceleration of the particle is constant.

    a Find the acceleration of the particle.

    b Find the position vector of the particle when its velocity is $(36\mathbf{i} + 12\mathbf{j})$ m s^{-1}.   [AQA (B) Jan 2001]

5 A ball is thrown vertically upwards from ground level with an initial speed of 7 m s^{-1}. Assume that no resistance forces act on the ball, so that it moves only under the influence of gravity.

    a Find the maximum height of the ball.

    b The ball hits the ground T seconds after it was thrown. Find T.   [AQA (B) Nov 2002]

3 Statics and forces

3.1 Resultant of forces

Finding the resultant force acting on a particle; finding the resultant of a number of forces acting at a point; modelling forces as vectors.

In Chapter 2, you were introduced to **resolving** vectors into their horizontal and vertical **components**. This practice can be applied to a **force**, as force is a vector quantity (it has both magnitude and direction). The exact definition of a force, measured in newtons (N), is given in Chapter 5. In this chapter, you will solve problems by resolving forces into their components.

Resultant of a horizontal and vertical force

If a force $\mathbf{F}$ N has horizontal component F_h N and vertical component F_v N, as shown in the diagram, then the magnitude of the resultant can be calculated using:

$$|\mathbf{F}| = \sqrt{(F_h^2 + F_v^2)}.$$

The angle made with the horizontal component can be calculated using

$$\tan \theta° = \frac{F_v}{F_h}.$$

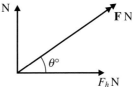

Recall:
You can also write this in vector form as:
$$\mathbf{F} = F_h\mathbf{i} + F_v\mathbf{j}.$$

Recall:
Another way of writing the magnitude of $\mathbf{F}$ is $F = |\mathbf{F}|$. This is also referred to as the modulus or size, F.

Example 3.1 A force of 8 N acts horizontally to the right and a force of 3 N acts vertically upwards. Find the magnitude of the resultant of these forces and the angle that the resultant makes with the horizontal force.

Step 1: Draw a force diagram resolving each force into its components.

Step 2: Find the resultant force.

$$|\mathbf{F}| = \sqrt{(F_h^2 + F_v^2)}$$
$$= \sqrt{8^2 + 3^2}$$

Step 3: Solve for unknowns.

$$= 8.54 \ldots$$

$$\tan \theta° = \frac{F_v}{F_h} = \frac{3}{8}$$

$$\theta = \tan^{-1}\left(\frac{3}{8}\right) = 20.6 \ldots$$

The magnitude of the resultant force is 8.54 N (3 s.f.) and the angle is 20.6° (3 s.f.).

Note:
You may be asked to find the angle that the resultant makes with the upward vertical. This is $90° - 20.6° = 69.4°$.

Example 3.2 Find the magnitude of the resultant and the angle that the resultant makes with the positive x-axis for the force $\mathbf{F} = (3\mathbf{i} - 4\mathbf{j})$ N, where $\mathbf{i}$ and $\mathbf{j}$ are perpendicular unit vectors.

Step 1: Draw a force diagram resolving each force into its components.

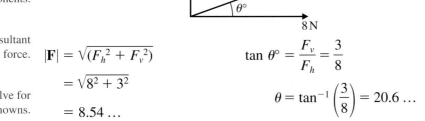

Recall:
$\mathbf{i} - \mathbf{j}$ notation.

Step 2: Find the resultant force.

$$|\mathbf{F}| = \sqrt{(F_h^2 + F_v^2)}$$
$$= \sqrt{(3^2 + 4^2)}$$

Step 3: Solve for unknowns.

$$= 5$$

The magnitude of the resultant force is 5 N.

$$\tan \theta° = \frac{|F_v|}{F_h} = \frac{4}{3}$$

$$\theta = \tan^{-1}\left(\frac{4}{3}\right) = 53.1 \ldots$$

The angle required is below the positive x-axis, so can be written as $-53.1°$ or $307°$ (3 s.f.).

Tip:
Use positive values to find an angle first.

Recall:
Positive angles are measured anticlockwise from the positive x-axis.

Resolving a force into components

If a force **F** N has magnitude F N and it acts at an angle $\theta°$ to the positive x-axis you can work out its horizontal and vertical components using simple trigonometry:

$F_h = F \cos \theta°$

$F_v = F \sin \theta°$

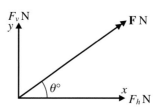

If **i** and **j** are perpendicular unit vectors then **F** can then be written in vector form as:

$\mathbf{F} = F \cos \theta° \mathbf{i} + F \sin \theta° \mathbf{j}$

Example 3.3 (In this question **i** and **j** are perpendicular unit vectors.)

a Find the horizontal and vertical components for a force of 10 N, which acts at an angle of 140° to the vector **i**.

b Write the force in the form $a\mathbf{i} + b\mathbf{j}$.

Step 1: Draw a force diagram resolving each force into its components.

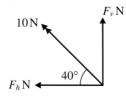

Step 2: Find the components in each direction.

a Horizontal component
= 10 cos 40° to the left
so $F_h = -7.66$ (3 s.f.)

Vertical component
= 10 sin 40° upwards
so $F_v = 6.43$ (3 s.f.)

b The required force is $(-7.66\mathbf{i} + 6.43\mathbf{j})$ N (3 s.f.).

Resultant of two or more forces

The resultant of two or more forces can be calculated by summing the vertical and horizontal components, separately.

Example 3.4 Find the resultant and the angle that the resultant makes with the 5 N force for the following system of forces.

Step 1: Draw the force diagram, and, where appropriate, redraw each force resolved into any two perpendicular directions.

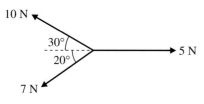

Step 2: Find the resultant in each direction.

Horizontal component:

$F_h = 5 - 10 \cos 30° - 7 \cos 20°$
$= -10.23...$

Vertical component:

$F_v = 10 \sin 30° - 7 \sin 20°$
$= 2.60...$

Step 3: Find the overall resultant and the angle.

$F = \sqrt{(F_h^2 + F_v^2)} = 10.5$ (3 s.f.)

$\theta = \tan^{-1}\left(\frac{F_v}{|F_h|}\right) = 14.3$ (3 s.f.)

The resultant is a force of 10.5 N at an angle of 166° with the 5 N force.

Example 3.5 The following forces, **P** N, **Q** N and **R** N, act on a particle, where **i** and **j** are unit vectors acting due east and due north respectively.

$$\mathbf{P} = (3\mathbf{i} + 4\mathbf{j}), \mathbf{Q} = (2\mathbf{i} - 5\mathbf{j}), \mathbf{R} = (-\mathbf{i} - 2\mathbf{j})$$

a Find the resultant force, **F**, in vector form.

b Also find the magnitude of the resultant and the angle that the resultant makes with the vector **i**.

Step 1: Find the resultant force by adding the **i** parts and the **j** parts separately.

a $\mathbf{F} = \mathbf{P} + \mathbf{Q} + \mathbf{R} = (3\mathbf{i} + 4\mathbf{j}) + (2\mathbf{i} - 5\mathbf{j}) + (-\mathbf{i} - 2\mathbf{j})$

$$= (3 + 2 - 1)\mathbf{i} + (4 - 5 - 2)\mathbf{j}$$

$$= (4\mathbf{i} - 3\mathbf{j})$$

Step 2: Calculate the magnitude and angle of the resultant force.

b $|\mathbf{F}| = \sqrt{(F_h^2 + F_v^2)} = \sqrt{((4)^2 + (-3)^2)} = 5$

$$\theta = \tan^{-1}\left(\frac{|F_v|}{F_h}\right) = \tan^{-1}\left(\frac{3}{4}\right) = 36.9 \text{ (3 s.f.)}$$

The angle that the force makes with the vector **i** is $-36.9°$ (3 s.f.).

> **Note:**
> There is no need to draw a vector diagram to resolve the components, as the forces are defined in **i**–**j** notation, where **i** and **j** are the horizontal and vertical unit vectors.

> **Note:**
> You can also use column vectors for this example.

> **Note:**
> You can also write this angle as $(360° - 36.9...°) = 323°$

3.2 Equilibrium of forces

Knowledge that the resultant force is zero if a body is in equilibrium.

A system of forces acting on a particle is in equilibrium when the resultant force is zero (there is no net force).

- **The algebraic sum of the horizontal components is 0 N.**
- **The algebraic sum of the vertical components is 0 N.**

Example 3.6 Find *P* and *Q* if the following system of forces is in equilibrium:

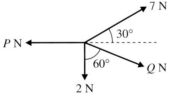

> **Recall:**
> $F_h = F\cos\theta°$ and $F_v = F\sin\theta°$ when $\theta°$ is defined as the angle between the vector and the positive *x*-axis. The angle that *Q* makes with the vertical is 60° so the angle *Q* makes with the horizontal is 30°.

Step 1: Draw the force diagram, and, where appropriate, redraw each of the forces resolved into any two perpendicular directions.

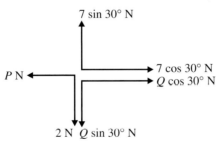

> **Note:**
> Define up and right as positive, down and left as negative.

Step 2: Find the resultant force in each direction (and equate each to zero when in equilibrium).

Step 3: Solve for unknowns.

Vertical components:

$7 \sin 30° - 2 - Q \sin 30° = 0$

$$Q = \frac{7 \sin 30° - 2}{\sin 30°} = 3$$

Force *Q* is 3 N.

Horizontal components:

$7 \cos 30° + Q \cos 30° - P = 0$

$P = 7 \cos 30° + 3 \cos 30°$

$= 8.660...$

Force *P* is 8.66 N (3 s.f.).

Example 3.7 The following diagram shows a system of forces acting on a particle in a plane. A third force is added so that the particle rests in equilibrium. Find the magnitude of this force and the angle that it makes with the horizontal.

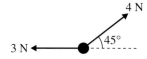

Step 1: Draw the force diagram, and, where appropriate, redraw each of the forces resolved into any two perpendicular directions.

Let the force which is added to keep the system in equilibrium be F N and let the angle it makes below the horizontal be $\theta°$.

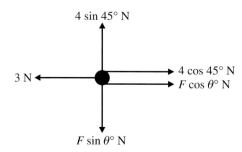

Note:
Include the unknown force and take a reasonable guess as to which direction it acts.

Step 2: Find the resultant force in each direction (and equate each to zero when in equilibrium).

Horizontal components:

$4 \cos 45° + F \cos \theta° - 3 = 0$

$F \cos \theta° = 3 - 4 \cos 45°$ ①

Vertical components:

$4 \sin 45° - F \sin \theta° = 0$

$F \sin \theta° = 4 \sin 45°$ ②

Step 3: Solve for unknowns.

Divide ② by ①

$$\tan \theta° = \frac{\sin \theta°}{\cos \theta°} = \frac{4 \sin 45°}{3 - 4 \cos 45°} = 16.48\ldots$$

$$\theta = \tan^{-1} 16.48\ldots$$

$$= 86.5 \text{ (3 s.f.)}$$

$$F = \frac{4 \sin 45°}{\sin 86.5°}$$

$$= 2.83 \text{ (3 s.f.)}$$

Recall:
From C2 that $\dfrac{\sin \theta}{\cos \theta} = \tan \theta$.

The added force is 2.83 N, acting at an angle of 86.5° below the horizontal.

When forces, defined in **i**–**j** notation, are in equilibrium then the resultant force is written as $(0\mathbf{i} + 0\mathbf{j})$ N $= \begin{bmatrix} 0 \\ 0 \end{bmatrix}$ N.

Example 3.8 The forces $\mathbf{P} = \begin{bmatrix} a \\ 4 \end{bmatrix}$, $\mathbf{Q} = \begin{bmatrix} 2 \\ -5 \end{bmatrix}$ and $\mathbf{R} = \begin{bmatrix} 1 \\ b \end{bmatrix}$, in newtons, act on a particle. The particle is at rest in equilibrium. Find a and b.

Step 1: Find the resultant force by adding the **i** parts and the **j** parts separately (and equate to $0\mathbf{i} + 0\mathbf{j}$ when in equilibrium).

$$\mathbf{P} + \mathbf{Q} + \mathbf{R} = \begin{bmatrix} a \\ 4 \end{bmatrix} + \begin{bmatrix} 2 \\ -5 \end{bmatrix} + \begin{bmatrix} 1 \\ b \end{bmatrix} = \begin{bmatrix} 0 \\ 0 \end{bmatrix}$$

$$\begin{bmatrix} a + 2 + 1 \\ 4 - 5 + b \end{bmatrix} = \begin{bmatrix} 0 \\ 0 \end{bmatrix}$$

$$\begin{bmatrix} a + 3 \\ b - 1 \end{bmatrix} = \begin{bmatrix} 0 \\ 0 \end{bmatrix}$$

Tip:
Sum the **i** and **j** terms separately.

Step 2: Solve for the unknowns by equating the **i** and **j** components separately.

Equating **i** components:

$a + 3 = 0$

$a = -3$

Equating **j** components:

$b - 1 = 0$

$b = 1$

In this exercise, where applicable, **i** and **j** are perpendicular unit vectors.

1 Find the magnitude of the resultant and the angle that the resultant makes with the positive *x*-axis for the following forces, giving your answers to one decimal place:

a 6 N ← 3 N **b** 5 N 5 N ← **c** 10 N 18 N **d** $(3\mathbf{i} + 2\mathbf{j})\,\text{N}$ **e** $(3\mathbf{i} - 5\mathbf{j})\,\text{N}$.

2 Find the horizontal and vertical components of the following forces:

   a a force of 10 N acting at 10° to the horizontal

   b a force of 25 N acting at 15° to the horizontal

   c a force of 35 N acting at 85° to the horizontal.

3 Find the magnitude of the horizontal and vertical components of the following forces that act on a particle, leaving your answer in the form $a\mathbf{i} + b\mathbf{j}$, where *a* and *b* are evaluated to one decimal place:

   a a force of 5 N acting at 120° to the horizontal

   b a force of 20 N acting at 30° below the horizontal

   c a force of 145 N acting at 220° to the horizontal.

 4 The diagram shows forces **P** N, **Q** N and **R** N acting on a particle.

The line of action of the force **Q** is in the horizontal direction, as shown in the diagram.

Find the magnitude of the resultant of these forces and the angle that the resultant makes with the direction of force **Q** if:

   a **P** = 10 acting at 30° above the horizontal, **Q** = 7 and **R** = 15 acting at 15° below the horizontal

   b **P** = 50 acting at 19° above the horizontal, **Q** = 45 and **R** = 100 acting at 18° below the horizontal.

 5 Find the magnitude of the resultant of the forces **P** N, **Q** N and **R** N and the angle that the resultant makes with the vector **i** if:

   a **P** = 7**i** + 2**j**, **Q** = 3**i** + 2**j**, **R** = 5**i** − 2**j**   **b** **P** = 3**i** − 2**j**, **Q** = −3**i** −5**j**, **R** = −2**i** + 2**j**.

6 The diagram shows the forces **A** N, **B** N and **C** N acting on a particle. The forces are in equilibrium.
Find **A** and **B** when:

   a $\theta = 25, \alpha = 40, \mathbf{C} = 10$

   b $\theta = 18, \alpha = 20, \mathbf{C} = 19$.

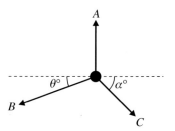

Quote your answers to a suitable degree of accuracy.

 7 A particle is acted on by a force of 15 N which acts on a bearing of 020°, and another force of 4 N which acts on a bearing of 230°. Find the magnitude of a third force which will keep the system in equilibrium, stating the angle of its line of action as a bearing.

8 A particle is in equilibrium at O under the action of forces **P** N, **Q** N and **R** N. Find the values of a and b if:

    a $\mathbf{P} = (a\mathbf{i} + 6\mathbf{j})$, $\mathbf{Q} = (5\mathbf{i} + b\mathbf{j})$, $\mathbf{R} = (9\mathbf{i} - 2\mathbf{j})$

    b $\mathbf{P} = (2\mathbf{i} - 5\mathbf{j})$, $\mathbf{Q} = (a\mathbf{i} - 5\mathbf{j})$, $\mathbf{R} = (-3\mathbf{i} + b\mathbf{j})$

    c $\mathbf{P} = (-4\mathbf{i} - 3\mathbf{j})$, $\mathbf{Q} = (2\mathbf{i} - 11\mathbf{j})$, $\mathbf{R} = (a\mathbf{i} + b\mathbf{j})$.

SKILLS CHECK **3A EXTRA** is on the CD

3.3 Types of force

Drawing force diagrams, identifying forces present and clearly labelling diagrams;
force of gravity; normal reaction forces; tensions in strings and rods.

Weight is the gravitational attraction between a particle of mass m kg and the Earth. It is often written as:

$$\text{weight} = mg$$

where g is the acceleration due to gravity and has approximate value 9.8 ms^{-2}. Weight is a force that always acts vertically downwards, towards the centre of the Earth.

Normal reaction force is the force exerted by a surface on a particle which lies in equilibrium. This force always acts in a direction that is perpendicular to the surface, opposing the direction of the weight.

Tension is the resisting force provided by a string, when the string holds the particle in equilibrium. It acts away from the particle.

Friction is the force that opposes motion on a **rough** surface. It is caused by the contact between the particle and the surface. On a **smooth** surface the friction is 0 N.

These forces are used in unit M1. You can apply the techniques learnt above to questions involving these forces acting on a particle.

Example 3.9   A block of mass 3 kg rests on a rough, horizontal surface. The block is pushed with a horizontal force of 3 N. The block is kept in equilibrium by a frictional force F N.

    a   Find F.

    b   Find the normal reaction.

Step 1: Draw the force diagram and, where appropriate, redraw each of the forces resolved into any two perpendicular directions.

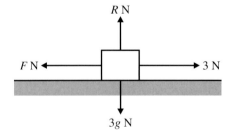

Note:
You can draw a pushing force on one side as a pulling force on the opposite side.

Step 2: Find the resultant force in each direction (and equate each to 0 when in equilibrium).

a   Horizontal components:

$$F - 3 = 0$$
$$F = 3$$

b   Vertical components:

$$R - 3g = 0$$
$$R = 3g$$
$$= 29.4$$

Recall:
The frictional force will act to oppose the pushing force of 3 N.

Note:
Equate the horizontal and vertical components to zero separately.

Step 3: Solve for unknowns. The frictional force is 3 N.

The normal reaction is 29.4 N.

Recall:
$g = 9.8$

Example 3.10 A particle of mass 3 kg rests on a smooth horizontal table. A string with tension T N acts an angle 45° above the horizontal and pulls the particle. Another force of 15 N acts at an angle of 30° below the horizontal and opposes the tension in the string to keep the particle in equilibrium. Find the reaction force and the tension in the string.

Note:
A rod can be treated in the same way as a string.

Step 1: Draw the force diagram and, where appropriate, redraw each of the forces resolved into any two perpendicular directions.

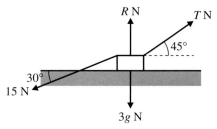

Note:
The method is the same as in the previous section. The only change is in describing the unknowns.

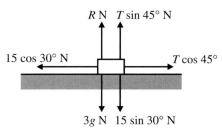

Step 2: Find the resultant force in each direction (and equate each to 0 when in equilibrium).

Step 3: Solve for unknowns.

Horizontal components:

$T \cos 45° - 15 \cos 30° = 0$

$T = \dfrac{15 \cos 30°}{\cos 45°} = 18.37\ldots$

The tension in the string is 18.4 N (3 s.f.).

Vertical components:

$R + T \sin 45° - 15 \sin 30° - 3g = 0$

$R = 15 \sin 30° + 3g - T \sin 45°$
$\quad = 23.90\ldots$

The normal reaction is 23.9 N (3 s.f.).

Sometimes a system of forces may act to suspend a particle freely in the air. In this case there will be no reaction force because the particle is not in contact with any surface.

Example 3.11 A particle of mass 2 kg is suspended freely by two strings and hangs in equilibrium as shown. Find the tension in the two strings.

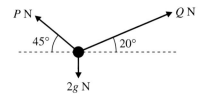

Step 1: Draw the force diagram and, where appropriate, redraw each of the forces resolved into any two perpendicular directions.

$P \sin 45°$ N $Q \sin 20°$ N

$P \cos 45°$ N $\longleftarrow$ $\longrightarrow$ $Q \cos 20°$ N

$2g$ N

Note:
Remember to include the weight.

Step 2: Find the resultant force in each direction (and equate each to 0 when in equilibrium).

Step 3: Solve for unknowns.

Horizontal components:

$Q \cos 20° - P \cos 45° = 0$

$Q = \dfrac{P \cos 45°}{\cos 20°}$

Substitute $P = 20.32\ldots$
$Q = 15.29\ldots$

Vertical components:

$P \sin 45° + Q \sin 20° - 2g = 0$

$2g = P \sin 45° + \dfrac{P \cos 45°}{\cos 20°} \sin 20°$

$19.6 = P(\sin 45° + \dfrac{\cos 45°}{\cos 20°} \sin 20)$
$P = 20.32\ldots$

The tensions in strings are 20.3 N (3 s.f.) and 15.3 N (3 s.f.).

Note:
Substitute for Q into the equation for the vertical component and solve for P. Then substitute the value for P back into the equation for Q.

Tip:
You need to practise solving these types of simultaneous equations.

3.4 Friction and the coefficient of friction

Friction, limiting friction, coefficient of friction and the relationship of $F \leqslant \mu R$.

When a particle of mass m kg rests on a rough surface, there is a frictional force **F** N which has a maximum value, F_{max} N. Suppose there is a pulling force of **P** N:

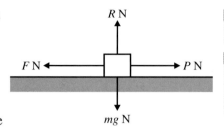

Consider the horizontal components of the forces; there are three situations that can arise:

1 If the force $P < F_{max}$ then the frictional force $F = P$, i.e. friction will take a value sufficient to maintain equilibrium.

2 If the force $P = F_{max}$ then the situation is said to be in **limiting equilibrium** and the system is at the point of moving.

3 If the $P > F_{max}$ then equilibrium is broken, the object will move and friction will maintain its maximum value, F_{max}.

When the situation is in limiting equilibrium then

$$F_{max} = \mu R$$

where μ is called the **coefficient of friction** and R is the normal reaction force of the particle. μ is a property of surface; rough surfaces have large values of μ and exert a large force of friction. Similarly, large values of the normal reaction cause large forces of friction.

Recall:
On a smooth surface, $\mathbf{F} = 0$ N.

Note:
If there is no pulling force then the frictional force exerted $= 0$ N.

Note:
If the frictional force acting is F then:
$$0 < F \leqslant F_{max}.$$

Recall:
The normal reaction force is related to the weight.

Example 3.12 A car of mass 1000 kg, is being pulled along a rough, horizontal surface by a rope. The tension in the rope is 5000 N and the rope is inclined at an angle of 15° above the horizontal. Given that the car is about to move, find the normal reaction force and the coefficient of friction.

Step 1: Draw the force diagram and, where appropriate, redraw each of the forces resolved into any two perpendicular directions.

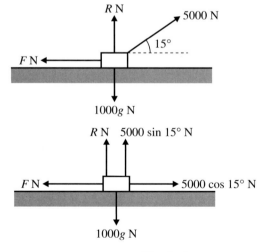

Note:
The method remains the same as in Sections 4.2 and 4.3; the only change is in the unknowns that you are asked to find.

Note:
$\mathbf{F} = \mu \mathbf{R}$ only in limiting equilibrium.

Note:
μ has no units.

Step 2: Find the resultant force in each direction (and equate each to 0 when in equilibrium).

Step 3: Solve for unknowns, using $F = \mu R$ when in limiting equilibrium.

Horizontal components:

$5000 \cos 15° - F = 0$

$F = 5000 \cos 15°$

Limiting equilibrium: $F = \mu R$

$$\mu = \frac{5000 \cos 15°}{8505.9\ldots}$$

The coefficient of friction is 0.57 (2 s.f.).

Vertical components:

$R + 5000 \sin 15° - 1000g = 0$

$R = 1000g - 5000 \sin 15°$

$\quad = 8505.9\ldots$

The normal reaction is 8500 N (2 s.f.).

Note:
Quote answers to 2 s.f. when using $g = 9.8$.

Note:
This is the smallest value of μ to maintain equilibrium. If μ is smaller, equilibrium will be broken so, for equilibrium to remain, $\mu \geqslant 0.57$ (2 s.f.).

Forces on an inclined plane

If a particle is on an inclined plane then, instead of resolving horizontally and vertically, resolve parallel to the plane and perpendicular to the plane, as this simplifies the algebra.

Parallel and perpendicular components of the weight

If a particle of mass m kg rests on a plane inclined at angle θ to the horizontal, the weight acts vertically downwards:

Draw the force diagram:

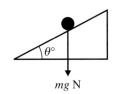

The components parallel and perpendicular to the plane are:

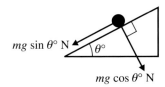

Note:
This can be seen by drawing a force triangle:

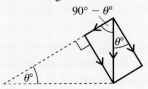

The parallel component of the weight is $mg \sin \theta°$ down the plane.
The perpendicular component of the weight is $mg \cos \theta°$.

Parallel and perpendicular components of the normal reaction force and friction

The normal reaction always acts perpendicular to the surface in contact so it only has a perpendicular component (R). Similarly, the friction is always parallel to the plane and so it only has a parallel component, in a direction opposing relative motion.

Note:
There is no need to resolve the normal reaction force and friction.

Example 3.13 A particle of mass 4 kg rests on a smooth plane inclined at 30° to the horizontal. A force of P N acts on the particle up the plane along the line of greatest slope. Find **a** the magnitude of the reaction force and **b** the magnitude of the force P.

Step 1: Draw the force diagram and, where appropriate, redraw each of the forces resolved into any two perpendicular directions.

Draw the force diagram:

The components parallel and perpendicular to the plane are:

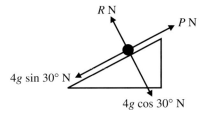

Note:
Resolve the forces in perpendicular directions (parallel and perpendicular to the plane).

Step 2: Find the resultant force in each direction (and equate each to 0 when in equilibrium).

Step 3: Solve for unknowns.

a Perpendicular to the plane:
$$R - 4g \cos 30° = 0$$
$$R = 4g \cos 30°$$
$$= 33.9\ldots$$

The reaction force has magnitude 34 N (2 s.f.).

b Parallel to the plane:
$$P - 4g \sin 30° = 0$$
$$P = 4g \sin 30°$$
$$= 19.6$$

The force P has magnitude 19.6 N.

Parallel and perpendicular components of a horizontal force

If a horizontal force **X** N acts on a particle on an inclined plane then:

- the component of **X** parallel to the plane is $X \cos \theta°$

- the component of **X** perpendicular to the plane is $X \sin \theta°$.

Draw the force diagram:

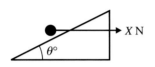

The parallel and perpendicular components are:

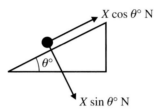

Example 3.14 A particle of mass 5 kg rests on a rough plane inclined at $\theta°$ to the horizontal where $\sin \theta° = 0.6$. The particle is kept in equilibrium by a horizontal force, of magnitude 40 N, acting in the vertical plane containing the line of greatest slope of the inclined plane through the particle. The particle is on the point of slipping up the plane.

a Find the force exerted by the plane on the particle.

b Find the coefficient of friction.

c How would the force diagram change if the particle were on the point of slipping down the plane?

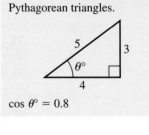
a Draw the force diagram:

The components parallel and perpendicular to the plane are:

Step 1: Draw the force diagram and, where appropriate, redraw each of the forces resolved into any two perpendicular directions.

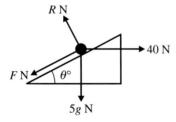

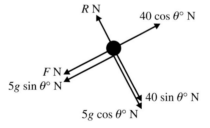

Step 2: Find the resultant force in each direction (and equate each to 0 when in equilibrium).

Perpendicular to the plane:
$$R - 40 \sin \theta° - 5g \cos \theta° = 0$$
Substitute for $\cos \theta°$ and $\sin \theta°$:
$$R - 24 - 4g = 0$$
$$R = 24 + 4g = 63.2$$
The normal reaction is 63 N (2 s.f.).

Parallel to the plane:
$$40 \cos \theta° - 5g \sin \theta° - F = 0$$

$$32 - 3g - F = 0$$
$$F = 2.6$$

Step 3: Solve for unknowns, using $F = \mu R$ when in limiting equilibrium.

b In limiting equilibrium, $F = \mu R$.

$$\mu = \frac{F}{R} = \frac{2.6}{63.2} = 0.04113\ldots$$

c If the particle was about to slip down the plane, the friction force would act up the plane, opposing the relative motion.

Parallel and perpendicular components of a force inclined at an angle $\alpha°$ to the plane

Draw the force diagram:

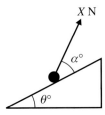

X N

$\alpha°$

$\theta°$

The components parallel and perpendicular to the plane are:

$X \sin \alpha°$ N

$X \cos \alpha°$ N

The component of **X** parallel to the plane is $X \cos \alpha°$ (up the plane).

The component of **X** perpendicular to the plane is $X \sin \alpha°$.

Example 3.15 A child of mass 35 kg sits on a toboggan, which rests on a rough inclined plane of angle 45°. The toboggan is light. The child is being pulled upwards by a rope, which is inclined at an angle of 15° above the line of greatest slope of the plane.

The coefficient of friction between the toboggan and the plane is 0.25. Equilibrium is about to be broken by the child sliding down the plane. By treating the child and the toboggan as a single particle, find the tension in the rope and the magnitude of the frictional force.

Step 1: Draw the force diagram and, where appropriate, redraw each of the forces resolved into any two perpendicular directions.

Draw the force diagram:

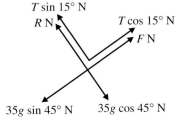

R N

T N

F N

15°

45°

$35g$ N

The components parallel and perpendicular to the plane are:

$T \sin 15°$ N

R N

$T \cos 15°$ N

F N

$35g \sin 45°$ N $35g \cos 45°$ N

Step 2: Find the resultant force in each direction (and equate each to 0 when in equilibrium).

Parallel to the plane:

$T \cos 15° + F - 35g \sin 45° = 0$

$F = 35g \sin 45° - T \cos 15°$

Perpendicular to the plane:

$T \sin 15° + R - 35g \cos 45° = 0$

$R = 35g \cos 45° - T \sin 15°$

Step 3: Solve for unknowns, using $F = \mu R$ when in limiting equilibrium.

In limiting equilibrium: $F = \mu R$

$35g \sin 45° - T \cos 15° = 0.25 (35g \cos 45° - T \sin 15°)$

$T(\cos 15° - 0.25 \sin 15°) = 35g \sin 45° - 0.25(35g \cos 45°)$

$T = 201.84\ldots$

$R = 190.29\ldots$

$F = \mu R = 0.25R = 47.57\ldots$

The tension in the rope is 200 N and the frictional force is 48 N (all 2 s.f.).

1 A particle of mass m kg rests on a rough horizontal plane. A horizontal force of P N acts on the particle. Find the magnitude of the normal reaction force and the frictional force if:

   a $m = 0.2, P = 7$ **b** $m = 3, P = 14$ **c** $m = 9, P = 24$

2 A particle of mass m kg lies at rest on a rough horizontal plane and is on the point of slipping. A string applies a tension T N at an angle $30°$ above the horizontal and pulls the particle along the plane. If the frictional force is F N along the plane and the normal reaction force of the plane on the particle is R N, find the magnitude of:

   a F and R if $m = 5$ and $T = 12$

   b T and m if $F = 12$ and $R = 2$

   c T and R if $F = 6$ and $m = 0.9$.

 3 Two men try to push a car of mass 1500 kg, which lies on a rough horizontal road. The two men apply forces of 50 N and 100 N. The car does not move. No other forces are present in the horizontal plane except friction.

   a Find the frictional force between the car and the road.

   b Given that the car is about to move find the coefficient of friction between the car and the road.

4 A particle of mass m kg is at rest on a rough horizontal table. A string with tension T N acting at an angle of $15°$ below the horizontal pulls the particle along the plane. The coefficient of friction between the plane and the block is μ and the normal reaction force exerted by the plane on the block is R N. Equilibrium is about to be broken; find the magnitude of:

   a R and μ if $m = 2$ and $T = 20$

   b T and R if $m = 0.5$ and $\mu = 0.25$

   c μ and m if $T = 15$ and $R = 15$.

5 A particle of mass 5 kg is suspended freely in the air by two light strings as shown in the diagram. Find the tensions in the two strings.

 6 The diagram shows a particle suspended from a horizontal beam by two unequal, light and inextensible strings. Given that the tension in the left string is 8 N and it makes an angle of $40°$ to the beam, and the other string makes an angle of $60°$ to the beam.

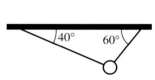

   a Find the tension in the other string.

   b Find the mass of the particle.

7 A particle of mass m kg lies on a rough plane, inclined at $\theta°$ to the horizontal where $\cos \theta° = \frac{12}{13}$. The system is in equilibrium.

   a Find **i** the magnitude of the normal reaction force and **ii** the frictional force, leaving your answer in terms of m and g, where g is the acceleration due to gravity.

   b Given also that the particle is in limiting equilibrium, find the coefficient of friction between the plane and the particle.

8 A particle of mass 3 kg lies on a rough plane inclined at $30°$ to the horizontal. A force of X N acts up the plane, along the line of greatest slope of the plane. The coefficient of friction between the particle and the plane is 0.5. Find the magnitude of X if:

   a the particle is on the point of slipping up the plane

   b the particle is on the point of slipping down the plane.

9 a Repeat question **8a** with the force of X N parallel to the plane replaced by a horizontal force of X N acting in the vertical plane containing the line of greatest slope of the inclined plane through the particle.

b Repeat question **8b** with the force of X N parallel to the plane replaced by a force of X N pulling the particle up the plane, acting at $45°$ above the line of greatest slope of the plane.

10 A car of mass 1500 kg is broken down on a rough plane inclined at an angle of $\theta°$ to the horizontal, where $\theta° = \sin^{-1}\left(\frac{7}{25}\right)$. It is being pulled up the plane by means of a towrope, which is acting at $9°$ above the line of greatest slope of the plane. The resistance between the plane and the car have magnitude 1000 N. The car is at rest in equilibrium and is about to move up the plane. Find **a** the tension in the towrope and **b** the magnitude of the force of the plane on the car.

SKILLS CHECK **3B EXTRA** is on the CD

Examination practice   Statics and forces

1 Two cables, AB and AC, are attached to a cable car, as shown in the diagram. The cable car has mass 450 kg.

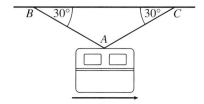

The cable car travels horizontally in the direction shown by the arrow. Model the cable car as a particle and assume that there is no air resistance present. As the cable car moves, the angles shown in the diagram do not change.

The cable car travels at a constant speed. Show that the tension in each cable is 4410 N.

[AQA (B) Jan 2002]

2 A load of mass 50 kg is supported, in equilibrium, by two ropes. One is at an angle of $30°$ to the vertical and the other is horizontal, as shown in the diagram. The tension in these ropes are T_1 newtons and T_2 newtons respectively.

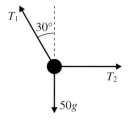

a Show that $T_1 = 566$, correct to 3 significant figures.

b Find T_2.

[AQA (B) Jun 2002]

3 Two forces $\mathbf{F}_1 = (3\mathbf{i} + 4\mathbf{j})$ N and $\mathbf{F}_2 = (6\mathbf{i} - 8\mathbf{j})$ N, act on a particle. The resultant of these two forces is $\mathbf{F}$. The unit vectors $\mathbf{i}$ and $\mathbf{j}$ are perpendicular.

a Find $\mathbf{F}$.

b Find the magnitude of $\mathbf{F}$.

c Find the acute angle between $\mathbf{F}$ and the unit vector $\mathbf{i}$.

[AQA (B) Jan 2003]

4

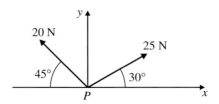

A particle P lies on a smooth horizontal surface. It is acted on by two horizontal forces of magnitudes 25 N and 20 N. Relative to horizontal axes, P_x and P_y, the directions of these two forces are as shown in the diagram. A third horizontal force **F** is required to keep P in equilibrium.

a Express the force of magnitude 25 N as a column vector, giving its components to one decimal place.

b Obtain **F** as a column vector, giving its components to one decimal place. [AQA (A) Jan 2002]

 5

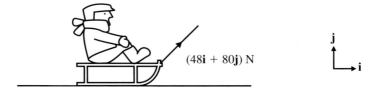

The diagram shows a child on a sledge which is being pulled at **constant speed** across a snowy horizontal field. The tension in the rope pulling the sledge is $(48\mathbf{i} + 80\mathbf{j})$ N, where the unit vectors **i** and **j** are horizontal and vertical, respectively, as shown in the diagram.

a Draw a diagram showing all the forces which act on the sledge.

b State the magnitude of the frictional force acting on the sledge.

c The coefficient of friction between the sledge and the surface is 0.4. Show that the normal reaction force between the sledge and the field is of magnitude 120 N.

d Find the total weight of the child and the sledge. [AQA (A) Jan 2002]

4 Momentum

4.1 Momentum

Concept of momentum.

The momentum of a particle of mass m kg, travelling with velocity $\mathbf{v}$ m s^{-1}, has magnitude mv. Momentum is a vector quantity and is measured in units of N s (newton seconds).

$$\text{Momentum} = m\mathbf{v} \text{ N s}$$

If the particle has speed v m s^{-1} then the momentum is given by:

$$\text{Momentum} = mv \text{ N s}$$

Recall:
$1\,\text{N} = 1\,\text{kg}\,\text{m}\,\text{s}^{-2}$.

Example 4.1 A particle of mass 3 kg is travelling with velocity of 4 m s^{-1}. Calculate:

  a the momentum of the particle

  b the change in momentum if the velocity of the particle increases to 8 m s^{-1}.

Step 1: Calculate unknowns using the definition of momentum.

  a $mv = 3 \times 4 = 12$
    The initial momentum of the particle is 12 N s.

  b $mv = 3 \times 8 = 24$
    The final momentum is 24 N s.
    The change in momentum $= 24 - 12 = 12$ N s.

Note:
You must consider the direction of momentum.

Example 4.2 A particle of mass 3 kg has initial velocity of 2.5 m s^{-1}. Find the change in momentum of the particle if:

  a the final velocity of the particle is 6 m s^{-1}

  b the final velocity of the particle is -6 m s^{-1}.

Note:
A negative sign indicates opposite direction.

Step 1: Draw a diagram of the initial and final situations

a Initial motion: Final motion:

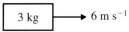

    3 kg → 2.5 m s^{-1} 3 kg → 6 m s^{-1}

Step 2: Calculate unknowns using the definition of momentum.

Initial momentum $= mv$ Final momentum $= mv$
               $= 3 \times 2.5$ $= 3 \times 6$
               $= 7.5$ $= 18$

Change in momentum $= 18 - 7.5 = 10.5$ N s.

Step 1: Draw a diagram of the initial and final situations

b Initial motion: Final motion:

    3 kg → 2.5 m s^{-1} 6 m s^{-1} ← 3 kg

Note:
The direction is important. The velocity to the right is taken as positive.

Step 2: Calculate unknowns using the definition of momentum.

Initial momentum $= mv$ Final momentum $= mv$
               $= 3 \times 2.5$ $= 3 \times -6$
               $= 7.5$ $= -18$

Change in momentum $= (7.5 - (-18))$ N s $= 25.5$ N s.

The principle of conservation of momentum applied to two particles.

Momentum is conserved between two colliding particles when there are no external forces acting. That is, **the total momentum before the collision equals the total momentum after the collision.** You can use this relation to solve problems involving colliding particles.

Example 4.3 A particle of mass 4 kg travels horizontally with a velocity of 8 m s^{-1}. It collides with another particle of mass 3 kg, which is at rest. After the impact the 4 kg mass has a velocity of -5 m s^{-1}.

Find the velocity of the 3 kg mass after the impact.

Note:
Let the speed of the 3 kg mass after the collision be v m s^{-1}.

Step 1: Draw a diagram for the initial and final situations.

Before collision: After collision:

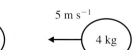

Note:
The direction of motion is very important, so choose which direction is positive at the start and stick with it.

Step 2: Calculate the total momentum before and after the collision.

Momentum before collision
$$= 4 \times 8 + 0$$
$$= 32$$

Momentum after collision
$$= 4 \times (-5) + 3v$$
$$= -20 + 3v$$

Note:
Let motion from left to right be positive.

Step 3: Calculate unknowns using the principle of conservation of momentum.

By the principle of conservation of momentum
$$32 = -20 + 3v$$
$$v = \frac{52}{3} = 17.33\ldots$$

The final velocity of the particle is 17 m s^{-1} (2 s.f.).

Recall:
Momentum before = momentum after.

Example 4.4 Two particles A and B of masses 2 kg and 1 kg respectively are moving towards each other in the same straight line with speeds $2u$ m s^{-1} and u m s^{-1}, respectively.

After the impact, the particles coalesce and continue to travel in the direction of particle A before impact, with speed 5 m s^{-1}.

Find the initial speeds of A and B.

Note:
When two particles coalesce, they join together and you can treat them as a single particle.

Step 1: Draw a diagram for the initial and final situations.

Before collision: After collision:

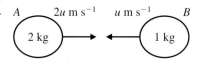   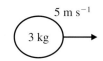

Note:
Motion from left to right is positive.

Step 2: Calculate the total momentum before and after the collision.

Momentum before collision
$$= 2(2u) - u = 3u$$

Momentum after collision
$$= 15$$

Tip:
It is useful to simplify each equation at this stage of the calculation.

Step 3: Calculate unknowns using the principle of conservation of momentum.

By the principle of conservation of momentum
$$3u = 15 \Rightarrow u = 5$$

The initial speeds of A and B are 10 m s^{-1} and 5 m s^{-1}, respectively.

Gun firing a bullet

When a gun fires a bullet the initial momentum is 0 N s, assuming there is no initial movement. But on firing the bullet there is a **recoil** in the gun which is in the opposite direction to the motion of the bullet, according to the conservation of momentum.

Recall:
Initial momentum = final momentum.

Example 4.5 A bullet is fired by a gun, which is 3 kg heavier than the bullet. When a shot is fired, the bullet travels in a straight line with velocity 200 m s^{-1} and the gun recoils in the opposite direction with velocity 5 m s^{-1}.

a Find the mass of the bullet.

b Find the mass of the gun.

Step 1: Draw a diagram for the initial and final situations.

a

5 m s^{-1} 200 m s^{-1}

Step 2: Calculate the total momentum before and after the collision.

Momentum before firing
= 0 N s

Momentum after firing
= $(3 + m)(-5) + 200m$
= $(195m - 15)$ N s

Step 3: Calculate unknowns using the principle of conservation of momentum.

By the principle of conservation of momentum
$$0 = 195m - 15$$
$$m = 0.0769\ldots$$

So the mass of the bullet is 77 g (2 s.f.).

b The mass of the gun is 3 kg + 0.77 kg = 3.08 kg (3 s.f.).

Note:
Let the mass of the bullet be m kg. Then the mass of the gun is $(m + 3)$ kg.

Note:
Motion from left to right is positive.

Note:
Because the value of m is so small it is better to convert it to grams.
$$1 \text{ kg} = 1000 \text{ g}$$

Momentum with vectors

You can also apply the principle of conservation of momentum to velocities given in vector form – momentum is conserved in each direction, horizontally and vertically. So you can treat each direction separately. It is easier, however, to keep the velocities in vector form, equate the components of the vectors before and after the collision, and then solve for the unknowns.

Example 4.6 A particle A of mass 5 kg, travelling with velocity $\begin{bmatrix} 3 \\ -2 \end{bmatrix}$ m s^{-1}, strikes a particle B of mass m kg, which is initially at rest. After the collision, A has velocity $\begin{bmatrix} 2 \\ -3 \end{bmatrix}$ m s^{-1} and B has velocity $\begin{bmatrix} 2 \\ v \end{bmatrix}$ m s^{-1}.

a Find m and v.
After the collision, particle B continues to travel and collides with another particle C of mass 1.5 kg, travelling with velocity $\begin{bmatrix} -4 \\ 6 \end{bmatrix}$ m s^{-1}. The particles coalesce after the collision.

b Find the velocity of the combined particles B and C after they collide.

c Find the speed of the combined particles B and C after the collision.

Step 1: Calculate the total momentum before and after the collision.

a Momentum before collision = $5\begin{bmatrix} 3 \\ -2 \end{bmatrix} + m\begin{bmatrix} 0 \\ 0 \end{bmatrix} = \begin{bmatrix} 15 \\ -10 \end{bmatrix}$ N s

Momentum after collision = $5\begin{bmatrix} 2 \\ -3 \end{bmatrix} + m\begin{bmatrix} 2 \\ v \end{bmatrix} = \begin{bmatrix} 10 + 2m \\ mv - 15 \end{bmatrix}$ N s

Note:
It is not necessary to draw a diagram here so, this step has been omitted. The other steps are the same.

Recall:
When adding vectors, add the components separately.

Step 2: Use principle of
conservation of momentum
(equate the **i**-parts and
j-parts) and solve for
unknowns.

By the principle of conservation of momentum

$$\begin{bmatrix} 15 \\ -10 \end{bmatrix} = \begin{bmatrix} 10 + 2m \\ mv - 15 \end{bmatrix}$$

i: $15 = 10 + 2m$
$m = 2.5$

j: $-10 = mv - 15$
$mv = 5$
$v = 2$

Step 1: Calculate the total
momentum before and
after the collision.

b Momentum before collision $= 2.5\begin{bmatrix} 2 \\ 2 \end{bmatrix} + 1.5\begin{bmatrix} -4 \\ 6 \end{bmatrix} = \begin{bmatrix} -1 \\ 14 \end{bmatrix}$ N s

Momentum after collision $= 4\begin{bmatrix} v_1 \\ v_2 \end{bmatrix}$ N s

Note:
Let the velocity of the
combined particles after the
impact be $\begin{bmatrix} v_1 \\ v_2 \end{bmatrix}$ m s-1.

Step 2: Use principle of
conservation of momentum
(equate the **i**-parts and
j-parts) and solve for
unknowns.

By the principle of conservation of momentum

$$\begin{bmatrix} -1 \\ 14 \end{bmatrix} = 4\begin{bmatrix} v_1 \\ v_2 \end{bmatrix}$$

i: $-1 = 4v_1$
$v_1 = -0.25$

j: $14 = 4v_2$
$v_2 = 3.5$

So the velocity of the combined particles B and C after the collision is
$\begin{bmatrix} -0.25 \\ 3.5 \end{bmatrix}$ m s^{-1}.

c The speed of the particles after the collision is
$\sqrt{((-0.25)^2 + 3.5^2)} = 3.5$ m s^{-1} (2 s.f.).

SKILLS CHECK **4A: Momentum**

1 Calculate the momentum for a particle of mass m kg and travelling with velocity v m s^{-1} when:

a $m = 2, v = 4$

b $m = 0.1, v = 18$

c $m = \frac{2}{3}, v = 15$.

2 Find the magnitude of the change in momentum of a particle of mass 8 kg that changes its speed from 2 m s^{-1} to:

a 14 m s^{-1} in the same direction

b 14 m s^{-1} in the opposite direction.

3 In parts **a**, **b** and **c**, two particles collide as shown. The diagrams show the situation before and after the impact. Find the unknown initial velocity u or the unknown final velocity v as appropriate:

a Before collision: After collision:

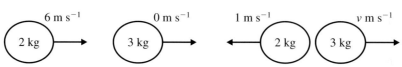

b Before collision:

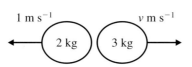

After collision:

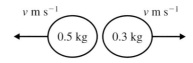

c Before collision:

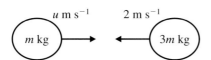

After collision:

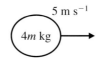

4 A car of mass 800 kg travelling on a smooth road with speed 6 m s^{-1} collides and coalesces with a stationary car of mass 500 kg. Find the speed of the combined cars after the collision.

 5 A gun of mass 3 kg fires a bullet of mass 20 g.

a If the gun recoils at 2 m s^{-1}, find the speed of the bullet after the shot is fired.

The bullet drives into a wall with this speed and stops after it has driven 2 cm into the wall in a horizontal direction.

b Find the magnitude of the resistance provided by the wall.

6 A particle P of mass 3 kg, travelling with velocity $\begin{bmatrix} 7 \\ 2 \end{bmatrix}$ m s^{-1}, strikes a particle Q of mass 2 kg, which is initially at rest. After the collision P has velocity $\begin{bmatrix} 4 \\ 1 \end{bmatrix}$ m s^{-1}.

a Find the velocity of Q after the collision.

b Hence find the speed of Q after the collision.

 7 Two identical particles A and B travel towards each other with velocities $(6\mathbf{i} - \mathbf{j})$ m s^{-1} and $(-2\mathbf{i} + 6\mathbf{j})$ m s^{-1} respectively, where $\mathbf{i}$ and $\mathbf{j}$ are perpendicular unit vectors. After they collide, the velocity of A is $(\mathbf{i} + \mathbf{j})$ m s^{-1}.

a Find the velocity of B after the collision.

b Hence find **i** the speed of B after the collision and **ii** the angle that the motion of B makes with the vector.

SKILLS CHECK **4A EXTRA** is on the CD

Examination practice **Momentum**

1 Two particles, A and B, of masses $2m$ kg and m kg respectively, are moving directly **towards** each other on a smooth horizontal surface. The speeds of A and B are 2 m s^{-1} and 6 m s^{-1} respectively.

The particles A and B collide and subsequently move directly **away** from each other with speeds $3V$ m s^{-1} and V m s^{-1} respectively.

Find the value of V.

[AQA (A) Jan 2003]

2 A bullet, of mass 50 grams, is fired from a gun, of mass 5 kg. The bullet hits a wooden block, of mass 3 kg, that is initially at rest on a rough, horizontal surface. The bullet becomes embedded in the block. The combined block and bullet slide 0.5 m before coming to rest. The coefficient of friction between the block and the surface is 0.8. Assume that the bullet always travels horizontally.

   a Show that the speed of the bullet before it hits the block is 170.8 m s^{-1}.

   b Find the speed at which the gun recoils after firing.

   c State **two** important assumptions, other than that the bullet travels horizontally, that you made to obtain your solutions.   [AQA (B) Jan 2001]

3 A particle A of mass 0.3 kg is moving with velocity $\begin{bmatrix} 7 \\ 4 \end{bmatrix}$ m s^{-1} when it collides with a stationary particle, B, of mass 0.5 kg. Immediately after the collision, B moves with velocity $\begin{bmatrix} 6 \\ 0 \end{bmatrix}$ m s^{-1}.

   a Find the velocity of A immediately after the collision.

   b Find the speed of A immediately after the collision.

   c State which of A and B moves faster after the collision.   [AQA (B) Jun 2002]

 4 A particle A, of mass 0.1 kg, is moving with velocity $\begin{bmatrix} 2 \\ 5 \end{bmatrix}$ m s^{-1} when it collides with another particle B, of mass m kg, which is moving with velocity $\begin{bmatrix} -1 \\ 0 \end{bmatrix}$ m s^{-1}. After the collision, A and B move with velocities $\begin{bmatrix} 1 \\ c \end{bmatrix}$ m s^{-1} and $\begin{bmatrix} 3 \\ 4 \end{bmatrix}$ m s^{-1} respectively.

   a Find the value of of m.

   b Find the value of of c.   [AQA (A) Nov 2002]

 5 Three supermarket trolleys, each of mass 20 kg, are placed on a straight line. As these trolleys move they are not subject to any resistance to motion. The first one is set in motion so that it moves at a speed of 4.5 m s^{-1} towards the second trolley. It then collides with the second trolley. After this collision the two trolleys continue to move together along the straight line at constant speed until they collide with the third trolley. After this collision all three trolleys move together along the straight line.

   a Show that the speed of the two moving trolleys after the first collision is 2.25 m s^{-1}.

   b Find the speed of the trolleys after the second collision.

   c After the second collision the combined trolleys are subject to a single, horizontal resistance force of magnitude 30 N.

      i Calculate the acceleration of the trolleys while this force acts.

      ii Find the distance, that the trolleys move after the second collision, before they come to rest.   [AQA (B) Jan 2002]

6 A particle P has mass 5 kg. It is moving along a straight line with speed 4 m s^{-1}, when it collides directly with another particle Q which is at rest. The mass of Q is m kg.

After the first collision P moves with a speed of 1.2 m s^{-1} and Q moves with a speed of 1.4 m s^{-1}.

   a If P and Q both move in the same direction after the collision, show that $m = 10$.

   b If P and Q move in opposite directions after the collision, find m.   [AQA (B) Jun 2002]

5 Newton's laws of motion

5.1 Newton's laws of motion and application of $F = ma$

Newton's laws of motion; simple application to the linear motion of a particle of constant mass.

Definition of a force

A force acting on an object causes the object to accelerate. The unit of force is the **newton** (N). A force of 1 N acting on a particle of mass 1 kg causes it to accelerate at 1 m s^{-2}.

Newton's first law states that a particle will remain at rest or will continue to move with constant speed in one direction unless acted on by an external (resultant) force. In other words, a change in velocity of an object is caused by the action of a resultant force on the particle, otherwise it remains at rest or maintains constant velocity.

Newton's second law states that the resultant force, F N, produces an acceleration, **a**, that is proportional to the resultant force according to:

$$\mathbf{F} = m\mathbf{a}$$

where m is the mass of the particle.

Note:
Learn this equation; remember **F** is the resultant force.

Newton's third law states that every action has an equal and opposite reaction, i.e. if one particle applies a force on another particle, the other particle applies an equal force on the first but in the opposite direction. This is the principle behind the normal reaction force, **R** (the surface exerts an equal and opposite force on the particle that rests on the surface).

Recall:
The normal reaction R is equal to the weight of a particle, mg, on a horizontal surface when no other forces act with a vertical component.

$\mathbf{F} = m\mathbf{a}$ describes motion in all possible directions. The particle will accelerate in the direction of the resultant force.

Horizontal motion

When the force acts horizontally on an object it will accelerate in the horizontal direction according to $\mathbf{F} = m\mathbf{a}$.

Example 5.1 A particle of mass 3 kg lies on a smooth horizontal plane. A horizontal force of 18 N acts on the particle. Calculate the acceleration of the particle.

Step 1: Draw the force diagram resolving the forces into any two perpendicular directions.

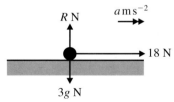

Note:
You do not need to resolve vertically in this example.

Step 2: Find the resultant force in each direction (and equate each to 0 if in equilibrium or to ma if accelerating).

Step 3: Solve for unknowns.

Horizontal components:

$F_h = ma$

$18 = 3a$

$a = 6$

Vertical components:

The particle is in equilibrium.

The particle accelerates at 6 m s^{-2} in the direction of the 18 N force.

47

Example 5.2 A particle of mass 5 kg is being pulled along a rough horizontal plane by a horizontal force of magnitude 15 N against a constant frictional force of magnitude 10 N. Given that the particle is initially at rest find:

a the acceleration of the particle

b the distance travelled by the particle in the first 3 seconds.

Step 1: Draw the force diagram (resolving the forces into any two perpendicular directions where necessary).

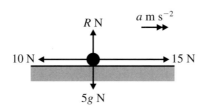

Note:
Consider forces acting up and to the right as positive and those acting down and to the left as negative.

Step 2: Find the resultant force in each direction (and equate each to 0 if in equilibrium or to *ma* if accelerating).

Step 3: Solve for unknowns.

a Horizontal components:

$$F_h = ma$$
$$15 - 10 = 5a$$
$$a = 1$$

Vertical components:

The particle is in equilibrium.

Note:
No net vertical force hence no vertical acceleration.

The particle accelerates at 1 m s^{-2} in the direction of the 15 N force.

b $s = ut + \frac{1}{2}at^2$

$s = 0(3) + \frac{1}{2}(1)(3)^2$

$s = 4.5$

The particle travels 4.5 m in 3 s.

Recall:
Chapter 3 on kinematics.
$s = ?$
$u = 0$
$v =$ not required
$a = 1$
$t = 3$

5.2 Friction and F = ma

Use of $F = \mu R$ as a model for dynamic friction.

Example 5.3 A particle of mass 2 kg is being pulled along a rough horizontal plane by a horizontal force of magnitude 12 N. It accelerates uniformly from rest to a speed of 2 m s^{-1} in 5 seconds. Find the acceleration of the particle and hence find the coefficient of friction between the plane and the particle.

Note:
Calculate the acceleration using the equations of motion before using the same method as previously.

Step 1: Draw the force diagram (resolving the forces into any two perpendicular directions where necessary).

$v = u + at$

$2 = 0 + a(5)$

$a = 0.4$

Recall:
The frictional force can be written as μR when equilibrium is broken.

Note:
The vertical forces are in equilibrium, so $F_v = 0$.

Step 2: Find the resultant force in each direction (and equate each to 0 if in equilibrium or to *ma* if accelerating).

Step 3: Solve for unknowns, using $F = \mu R$.

Horizontal components:

$F_h = ma$

$12 - F = 2(0.4)$

$F = 11.2$

$F = \mu R$

$\mu = \dfrac{F}{R} = \dfrac{11.2}{19.6} = 0.57$ (2 s.f.)

Vertical components:

$F_v = 0$

$R - 2g = 0$

$R = 19.6$

The normal reaction is 19.6 N.

Vertical motion

When the net force acts vertically on an object it will accelerate in the vertical plane according to **F** = *m***a**.

Example 5.4 A stone of mass 2 kg is attached to the lower end of a string hanging vertically. The particle is raised and moves with an acceleration of 5 m s^{-2}. Find the tension in the string.

Step 1: Draw the force diagram resolving the forces into any two perpendicular directions.

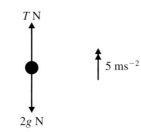

Note:
Let the tension in the string be T N.

Note:
Forces are all vertical and so do not need to be resolved into two directions.

Step 2: Find the resultant force in each direction (and equate each to 0 if in equilibrium or to ma if accelerating).

Step 3: Solve for unknowns.

No horizontal components.

Vertical components:
$$F_v = ma$$
$$T - 2g = 2(5)$$
$$T = 2g + 10$$
$$= 29.6$$

The tension in the string is 29.6 N.

Motion on an inclined plane

You can also apply $\mathbf{F} = m\mathbf{a}$ when a particle is on an inclined plane if there is a resultant force that will cause the particle to accelerate up or down the plane. The resultant force parallel to the plane will be referred to as $F_\parallel$.

Example 5.5 A particle of mass 6 kg is pulled up a smooth slope inclined at 45° to the horizontal, by means of a light, inextensible string which acts at an angle of 30° to the slope. The particle has an acceleration of 4 m s^{-2} up the slope. Show that the tension, T N, in the string is given by

$$T = (16 + 2g\sqrt{2})\sqrt{3}$$

where g is the acceleration due to gravity.

Step 1: Draw the force diagram resolving the forces into any two perpendicular directions.

Drawing a force diagram:

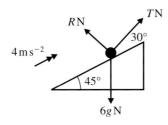

The components parallel and perpendicular to the plane are:

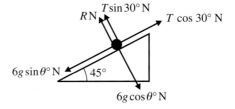

Recall:
$$\sin 45° = \frac{\sqrt{2}}{2}$$
$$\cos 30° = \frac{\sqrt{3}}{2}$$

Step 2: Find the resultant force in each direction (and equate each to 0 if in equilibrium or to ma if accelerating).

Step 3: Solve for unknowns.

Parallel to the plane:
$$F_\parallel = ma$$
$$T\cos 30° - 6g\sin 45° = 6(4)$$
$$T\frac{\sqrt{3}}{2} - 3g\sqrt{2} = 24$$

You do not need to resolve perpendicular to the plane.

Note:
Multiplying by $\sqrt{3}$ is one way of removing the square root with T.

Multiplying by 2: $\quad T\sqrt{3} - 6g\sqrt{3} = 48$
$$T\sqrt{3} = 48 + 6g\sqrt{2}$$

Multiplying by $\sqrt{3}$: $\qquad 3T = (48 + 6g\sqrt{2})\sqrt{3}$
$$T = (16 + 2g\sqrt{2})\sqrt{3} \quad \text{as required.}$$

Example 5.6 A block of mass 13 kg is released from rest on a rough plane inclined at an angle of $\theta°$, where $\sin \theta° = \frac{5}{13}$. It slides down the plane and reaches a speed of 4 m s^{-1} in 2 seconds. Using the equations of motion, find the coefficient of friction between the plane and the block.

Drawing a force diagram:

The components parallel and perpendicular to the plane are:

Step 1: Draw the force diagram resolving the forces into any two perpendicular directions.

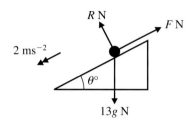

Note:
Use $v = u + at$ with $v = 4$, $u = 0$, $t = 2$.
$4 = 0 + 2a$
$a = 2$
The acceleration is 2 m s^{-2}.

Step 2: Find the resultant force in each direction (and equate each to 0 if in equilibrium or to ma if accelerating).

Parallel to the plane:

$$F_{\parallel} = ma$$
$$13g \sin \theta° - F = 13(2)$$
$$5g - F = 26$$

Step 3: Solve for unknowns, using $F = \mu R$.

$$F = 5g - 26$$
$$= 23$$
$$F = \mu R$$
$$\mu = \frac{F}{R} = \frac{23}{12g} = 0.20...$$

Perpendicular to the plane:
(in equilibrium)

$$R - 13g \cos \theta° = 0$$
$$R - 12g = 0$$
$$R = 12g$$

Note:
$\sin \theta° = \frac{5}{13}$
$\cos \theta° = \frac{12}{13}$

Note:
The particle is in equilibrium perpendicular to the plane.

The coefficient of friction is 0.20 (2 s.f.).

Motion described in i–j notation

$\mathbf{F} = m\mathbf{a}$ can be applied to problems using $\mathbf{i}$–$\mathbf{j}$ vector notation.

Example 5.7 Forces of $(5\mathbf{i} + 4\mathbf{j})$ N, $(2\mathbf{i} - 5\mathbf{j})$ N, $(\mathbf{i} + 7\mathbf{j})$ N act on a particle of mass 5 kg, where $\mathbf{i}$ and $\mathbf{j}$ are perpendicular unit vectors.

a Find the magnitude of the acceleration.

b Find the direction in which the particle is accelerating.

Recall:
$\mathbf{i}$–$\mathbf{j}$ notation in Chapter 2.

a $\mathbf{F} = m\mathbf{a}$

Step 1: Find the resultant force by adding the $\mathbf{i}$ parts and the $\mathbf{j}$ parts separately (and equate to $0\mathbf{i} + 0\mathbf{j}$ if in equilibrium, or to $m\mathbf{a}$ if accelerating).

$$(5\mathbf{i} + 4\mathbf{j}) + (2\mathbf{i} - 5\mathbf{j}) + (\mathbf{i} + 7\mathbf{j}) = 5\mathbf{a}$$
$$8\mathbf{i} + 6\mathbf{j} = 5\mathbf{a}$$
$$\mathbf{a} = \tfrac{1}{5}(8\mathbf{i} + 6\mathbf{j})$$
$$|\mathbf{a}| = \sqrt{\left(\frac{8}{5}\right)^2 + \left(\frac{6}{5}\right)^2} = 2$$

Note:
The method is similar to previous examples, but you do not need to draw a force diagram.

Step 2: Solve for unknowns.

The magnitude of the acceleration is 2 m s^{-2}.

b $$\tan \theta° = \frac{\frac{6}{5}}{\frac{8}{5}} = \frac{3}{4}$$
$$\theta° = 36.9° \text{ (3 s.f.)}$$

Note:
Let θ be the angle between the vector and the x-axis.

The direction in which the particle is accelerating is at an angle of $36.9°$ to the vector $\mathbf{i}$.

1 Find the acceleration produced when a particle of mass m kg is acted on by a resultant horizontal force of F N when:

   a $F = 8, m = 2$ **b** $F = 12, m = 6$ **c** $F = 0.5, m = 0.1$ **d** $F = \frac{1}{2}, m = 0.9$.

2 Find the resultant horizontal force that acts on a particle of mass m kg to produce an acceleration of a m s^{-2} when:

   a $m = 3, a = 2$ **b** $m = 0.55, a = 3$ **c** $m = 0.3, a = 0.9$.

3 A particle of mass 3 kg is being pulled across a rough horizontal plane by a horizontal force of 10 N. The coefficient of friction between the particle and the plane is 0.1.

   a Find the magnitude of the acceleration of the particle.

   b Given also that the particle is initially at rest, find the distance moved by the particle in the first 4 seconds.

4 A truck of mass 1500 kg is brought to rest in 5 seconds from a speed of 15 m s^{-1} on a smooth horizontal plane. Neglecting air resistance, find the braking force required to achieve this.

 5 The driving force produced by the engine of a car of mass one tonne causes a car to accelerate uniformly from rest to a speed of 8 m s^{-1} in 6 seconds along a rough horizontal road. If the coefficient of friction between the car and the plane is 0.3, find the driving force of the car.

6 A concrete block is raised vertically upwards, by a rope. It starts from rest and travels a vertical distance of 4 m in 6 seconds.

   a Given that the tension in the rope is 400 N, find the mass of the block.

   b What assumptions have you made when modelling this situation?

7 A boy of mass 30 kg slides down a smooth plane inclined 45° to the horizontal. Find the acceleration of the boy.

8 Repeat question **5**, with the car moving up a rough plane inclined 30° to the horizontal.

 9 A particle of mass 6.5 kg is in limiting equilibrium on a rough plane inclined at $\theta°$ to the horizontal where $\sin \theta° = \frac{5}{13}$.

   a Find the coefficient of friction between the plane and the particle.

   A horizontal force of 78 N is now applied to the particle so that it starts to accelerate up the plane.

   b Find the acceleration of the particle and hence find how far up the plane it moves in 2 seconds.

10 Forces of $(10\mathbf{i} + 2\mathbf{j})$ N, $(\mathbf{i} - 4\mathbf{j})$ N, $(2\mathbf{i} - 7\mathbf{j})$ N act on a particle of mass 5 kg.

   a Find the magnitude of the acceleration of the particle.

   b Find the angle that the acceleration makes with the vector $\mathbf{i}$.

11 The forces $\mathbf{P}$, $\mathbf{Q}$ and $\mathbf{R}$ (in newtons) act on a particle of mass 0.5 kg and produce an acceleration of $(\mathbf{i} - 4\mathbf{j})$ m s^{-2}.
$\mathbf{P} = (a\mathbf{i} - 4\mathbf{j})$, $\mathbf{Q} = (2\mathbf{i} + 4\mathbf{j})$, $\mathbf{R} = (-5\mathbf{i} + b\mathbf{j})$
Find the values of the constants a and b.

1

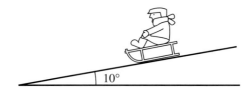

   a Alex is on her sledge, sliding down a snowy place as shown in the diagram. The slope is inclined at 10° to the horizontal and the coefficient of friction between the sledge and the slope is 0.12.

      The total mass of Alex and the sledge is 40 kg.
      i Show that the frictional force between the sledge and the slope is approximately 46 N.
      ii Find the acceleration of the sledge.

   b When the sledge reaches the bottom of the slope, Alex's brother pulls it back up the slope with Alex still sitting on it. He pulls the sledge at constant speed with a force which is parallel to the slope. Find the magnitude of this force.   [AQA (A) Nov 2002]

 2 A sledge, of mass 12 kg, is pulled up a rough slope which is inclined at an angle of 10° to the horizontal. The coefficient of friction between the slope and the sledge is 0.2.

   a The sledge is pulled by a rope that is parallel to the slope, as shown in the diagram.

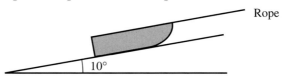

      i   Draw a diagram to show the forces acting on the sledge.
      ii Find the magnitude of the normal reaction force acting on the sledge.
      iii Given that the acceleration of the sledge is 0.5 m s^{-2}, show that the tension in the rope is approximately 50 N.

   b The sledge is then pulled with the rope at an angle of 30° to the slope, as shown in the diagram.

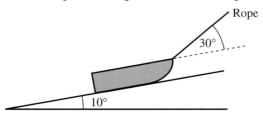

      Find the acceleration of the sledge if the tension in the rope is 60 N.

   c Write down **two** modelling assumptions that you have made.   [AQA (B) Nov 2002]

3 A ball is thrown vertically upwards from ground level. Throughout its motion it is acted on by gravity and a resistance force of constant magnitude.

   The ball reaches a maximum height of 1.5 metres after 0.5 seconds.

   a Find
      i the initial speed of the ball,
      ii its acceleration as it is moving upwards.

   The mass of the ball is 0.2 kg.

   b Show that the magnitude of the resistance force is 0.44 N.

   c Find the acceleration of the ball as it falls back to the ground.

   d Find the total time that the ball is in the air.   [AQA (B) Jun 2001]

4 A block, of mass 5 kg, is held at rest on a rough plane, which is inclined at 30° to the horizontal. The block is released and slides down the plane. The coefficient of friction between the block and the plane is 0.2.

   a Draw a diagram to show the forces acting on the block as it slides.

   b Show that the magnitude of the friction force acting on the block is approximately 8.5 N.

   c Find the acceleration of the block.

   d Find the speed of the block when it has travelled 1.2 metres down the slope.   [AQA (B) Jan 2003]

 5 Two constant forces $F_1 = (4i + 16j)$ N and $F_2 = (6i - 11j)$ N act on a particle. A force F_3 also acts on the particle. The mass of the particle is 8 kg and the unit vectors i and j are perpendicular.

   a In the case when the particle moves with a constant velocity, find F_3.

   b In the case when the acceleration of the particle is $(2i + 3j)$ m s^{-2}, find the magnitude of F_3.
                                                         [AQA (B) Jun 2001]

6 [In this question the value of g should be taken to be 9.8 m s^{-2}.]

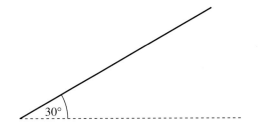

Kate and Joe are playing on a children's slide. The slide is straight and inclined at an angle of 30° to the horizontal, as shown in the diagram. Kate's mass is 25 kg and the coefficient of friction between Kate and the slide is 0.25.

   a Kate slides down the slide.

     i Show that the frictional force between her and the slide is approximately 53 N.

     ii Find her acceleration.

   b On one occasion, before Kate slides down, she is held at rest on the slide by Joe. He exerts a force on her which is parallel to the slide.

     i Find the minimum possible value of this force.

     ii Joe then releases Kate and she slides down. Find her speed after sliding 5 metres from her release point.   [AQA (A) Jun 2001]

6 Connected particles

6.1 Connected particles

Connected particle problems.

Horizontal motion of connected particles

When a particle is connected to another particle by a string then you can analyse the motion of the two **connected** particles by using Newton's third law. Both particles will experience the same magnitude of force, T N, but in opposite directions.

To analyse this motion apply $\mathbf{F} = m\mathbf{a}$ to each particle, separately.

Note:
Because they are connected they will both have the same acceleration.

Recall:
What assumptions must be made?

Example 6.1 Two particles of masses 5 kg and 7 kg are connected by an inextensible string. The particle of mass 7 kg is being pulled by a horizontal force of 70 N along a rough, horizontal surface. Given that the coefficient of friction between each particle and the surface is 0.25, find the acceleration of the system and the tension in the string. What assumption have you made about the string?

Note:
Define the direction of acceleration as positive.

Step 1: Draw the force diagram resolving the forces into any two perpendicular directions.

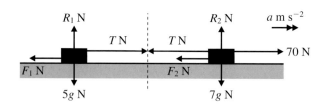

Note:
Each particle will have a different normal reaction because their masses are different.

Step 2: Find the resultant force in each direction (and equate to 0 if in equilibrium, to ma if accelerating)

5 kg particle:

$$F_v = 0 \text{ (in equilibrium)}$$
$$R_1 - 5g = 0 \qquad \text{①}$$
$$F_h = ma$$
$$T - F_1 = 5a \qquad \text{②}$$

Step 3: Solve for unknowns, using $F = \mu R$.

$$R_1 = 49 \quad \dots \text{from ①}$$
$$F_1 = \mu R_1 = 0.25(49)$$
$$= 12.25$$

7 kg particle:

$$F_v = 0 \text{ (in equilibrium)}$$
$$R_2 - 7g = 0 \qquad \text{③}$$
$$F_h = ma$$
$$70 - T - F_2 = 7a \qquad \text{④}$$

$$R_2 = 68.6 \quad \dots \text{from ③}$$
$$F_2 = \mu R_2 = 0.25(68.6)$$
$$= 17.15$$

Note:
Treat each particle separately, as if separated by the broken line in the diagram.

Note:
There is no vertical acceleration (no net vertical force).

Substitute into ②

$$T - 12.25 = 5a \qquad \text{⑤}$$

Substitute into ④

$$70 - T - 17.15 = 7a \qquad \text{⑥}$$

Add equations ⑤ and ⑥: $\quad 40.6 = 12a$

$$a = 3.383\dots = 3.4 \text{ (2 s.f.)}$$
$$T = 5a + 12.25 \qquad \dots \text{from ⑤}$$
$$= 29.166\dots = 29.2 \text{ (2 s.f.)}$$

Note:
Add equations ⑤ and ⑥ to eliminate T and solve for a.

Both particles accelerate at 3.4 m s^{-2} with a tension of 29 N in the connecting string.

We have assumed that the string is light, otherwise there would have to be a vertical component for the weight of the string.

Vertical motion of connected particles

This method also applies to particles that are connected vertically.

Example 6.2 A light, inextensible string connects two bricks of equal mass 5 kg, one above the other. The system is lowered by a tow bar, which is attached to the topmost brick. It takes 10 seconds for the bricks to travel a vertical distance of 15 m, starting from rest.

    a Find the acceleration of the bricks.

    b Use this to find the tensions in the string and the tow bar.

Step 1: Draw the force diagram resolving the forces into any two perpendicular directions.

a Use $s = ut + \frac{1}{2}at^2$

$$15 = 0(10) + \frac{1}{2}a(10)^2$$

$$a = 0.3$$

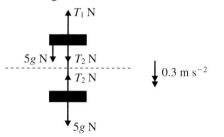

> **Note:**
> Let the tension in the tow bar be T_1, and the tension in the string be T_2.

Step 2: Find the resultant force in each direction (equate to 0 if in equilibrium, or to ma if accelerating).

Step 3: Solve for unknowns.

b Top brick:

$$F_v = ma$$
$$T_2 + 5g - T_1 = 5(0.3) \quad \textcircled{1}$$

No horizontal components.

Bottom brick:

$$F_v = ma$$
$$5g - T_2 = 5(0.3) \quad \textcircled{2}$$

No horizontal components.

$$T_2 = 5g - 1.5 \qquad \text{...from } \textcircled{2}$$
$$= 47.5$$
$$T_1 = T_2 + 5g - 1.5 \qquad \text{...from } \textcircled{1}$$
$$= 95$$

> **Note:**
> Define the direction of acceleration (down) as positive.

> **Note:**
> Alternatively equations ① and ② could be solved simultaneously.

The tension in the tow bar is 95 N and the tension in the string is 47.5 N.

Particles connected by pulleys

The motion of two particles that are connected by a light, inextensible string passing over a smooth fixed pulley can be similarly analysed. **F** = m**a** is applied *separately* to each particle. As the particles are connected by a light, inextensible string the magnitude of the acceleration is the same for both particles (but it acts in opposite directions). As the pulley is smooth, the tension in the string is the same throughout the string.

Example 6.3 Particles of mass m kg and $3m$ kg are connected by a light, inextensible string, which passes over a smooth fixed pulley.

    a Find, in terms of g, **i** the acceleration of the system and **ii** the force exerted on the pulley.

    b Assuming the particle does not reach the pulley when the system is released from rest, find the distance moved by one of the particles in 3 s.

Step 1: Draw the force diagram resolving the forces into any two perpendicular directions.

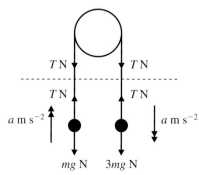

> **Note:**
> Remember to imagine a line half way down the string as shown.

> **Note:**
> Mass m moves up whilst mass $3m$ moves down.

Step 2: Find the resultant
force in each direction
(equate to 0 if in
equilibrium, or to *ma* if
accelerating).

Step 3: Solve for
unknowns.

a For mass *m*:

$$T - mg = ma \qquad \text{①}$$

No horizontal components.

For mass 3*m*:

$$3mg - T = 3ma \qquad \text{②}$$

No horizontal components.

①+②: $2mg = 4ma$

i

$$a = \frac{2g}{4} = \frac{g}{2}$$

The acceleration of the system is $\frac{g}{2}$ m s^{-2}.

ii The force exerted on the pulley is $2T$ N, where

$$T = ma + mg \qquad \text{...from ①}$$

$$= \frac{3mg}{2}$$

So, the force exerted on the pulley is $3mg$ N downwards.

b For the distance travelled in 3 seconds use $s = ut + \frac{1}{2}at^2$

$$s = 0(3) + \frac{1}{2}\left(\frac{g}{2}\right)3^2 = 22.1 \text{ (3 s.f.)}$$

So, the distance travelled is 22.1 m.

> **Recall:**
> Equations of motion in Chapter 2.

Horizontal and vertical motion of particles connected by pulleys

Sometimes a pulley can separate two particles, one of which is resting on a plane and the other hanging freely.

Example 6.4 Two particles *A* and *B* of masses 0.5 kg and 0.7 kg respectively are connected by a light, inextensible string. Particle *A* lies on a rough horizontal table 8 m from a smooth peg at the edge of the table. The string passes over the peg and particle *B* hangs freely 2 m from the ground. The coefficient of friction between particle *A* and the horizontal surface is 0.2. The system is released from rest. Find:
a the acceleration of the system
b the time taken for *B* to reach the ground
c the distance that *A* travels along the table, after *B* reaches the ground, before it comes to rest.

> **Note:**
> A smooth peg can be treated in exactly the same way as a smooth pulley.

Step 1: Draw the force
diagram resolving the
forces into any two
perpendicular directions.

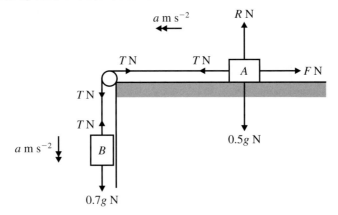

> **Note:**
> Apply $\mathbf{F} = m\mathbf{a}$ horizontally for particle *A* and vertically for particle *B*. Define the direction of motion as positive.

Step 2: Find the resultant
force in each direction
(equate to 0 if in
equilibrium, or to *ma* if
accelerating).

For 0.5 kg mass:
Horizontal motion:

$$F_h = ma$$
$$T - F = 0.5a \qquad \text{①}$$

Vertical motion:
$$F_v = 0 \text{ (in equilibrium)}$$
$$R - 0.5g = 0 \qquad \text{②}$$

For 0.7 kg mass:
No horizontal motion.

Vertical motion:
$$F_v = ma$$
$$0.7g - T = 0.7a \qquad \text{③}$$

> **Note:**
> Consider motion whilst particle *B* hangs freely.

Step 3: Solve for unknowns using $F = \mu R$.

a $R = 4.9$ ...from ②

$F = \mu R = 0.98$

Substitute into ① $T - 0.98 = 0.5a$ ④

$5.88 = 1.2a$

$a = 4.9$

The acceleration of the particles as B falls is 4.9 m s^{-2}.

b $s = ut + \frac{1}{2}at^2$ (where $s = 2, u = 0, a = 4.9, t = ?$)

$2 = 0(t) + \frac{1}{2}(4.9)t^2$

$t = \sqrt{\dfrac{4}{4.9}} = 0.903\ldots$

> **Note:**
> Use the value of a to calculate the time and final velocity while the string is taut.

Particle B falls for 0.90 s (2 s.f.) before hitting the ground.

Step 1: Find the final velocity of the particle while the string is taut.

c   Find the velocity of the particles as B hits the ground:

$v^2 = u^2 + 2as$ (where $s = 2, u = 0, v = ?, a = 4.9$)

$v^2 = 0(2) + 2(4.9)(2) = 19.6$

$v = 4.427\ldots$

Step 2: Find the new acceleration of the particle when there is no tension.

Find the acceleration of particle A:

You can apply $\mathbf{F} = m\mathbf{a}$ to particle A or you can recognise that the equation will be the same as the one you did earlier but without any tension. So, you can use equation (4) but remove the tension to get the new acceleration when the particles are no longer connected:

$-0.98 = 0.5a$ ...from ④

$a = -1.96$

> **Note:**
> Consider A after B hits the ground.

> **Note:**
> The negative sign means that the particle is decelerating.

Step 3: Use the equations of motion to find the further distance moved by the particle until it stops.

Find the further distance travelled by A:

$v^2 = u^2 + 2as$ (where $u = 4.427\ldots, v = 0, a = -1.96$)

$0^2 = 4.427\ldots^2 + 2(-1.96)s$

$s = 5$

> **Note:**
> The particle now decelerates from the velocity found in Step 1 to 0 m s^{-1} under this new acceleration.

Particle A travels 5 m along the table after particle B hits the ground (the total distance travelled by A is $(5 + 2) \text{ m} = 7 \text{ m}$).

Example 6.5   Two particles P and Q, of mass 10 kg and 15 kg respectively, are connected by a light, inextensible string which passes over a light, smooth pulley. Particle P rests on a smooth plane inclined at $\theta°$ to the horizontal, where $\sin \theta° = \frac{3}{5}$. Particle Q hangs vertically on the edge of the plane, 2 m above a horizontal plane.

a   Show that the acceleration of the system is 3.5 m s^{-2}.

b   Find the tension in the string.

c   Find the total distance that P travels up the plane, given that the string breaks after Q has travelled 1 m.

Step 1: Draw the force diagram resolving the forces into any two perpendicular directions.

Draw a force diagram:

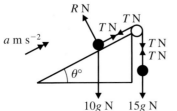

Resolve parallel to the plane for P and vertically for Q:

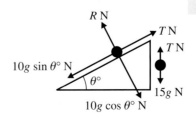

> **Note:**
> Define the direction of motion as positive.

Step 2: Find the resultant force in each direction (and equate to 0 if in equilibrium, to *ma* if accelerating).

Particle *P*:

Parallel to plane:

$$F_\| = ma:$$

$$T - 10g \sin \theta = 10a$$

$$T - 6g = 10a \qquad \text{①}$$

Particle *Q*:

Vertical motion:

$$F_v = ma$$

$$15g - T = 15a \qquad \text{②}$$

Tip: Use accurate values in calculations to avoid rounding errors.

Step 3: Solve for unknowns

a $9g = 25a \Rightarrow a = \dfrac{9 \times 9.8}{25} = 3.528$

The acceleration of the system is 3.5 m s^{-2} (2 s.f.).

Note: Add ① and ②.

b $T = 10a + 6g = 94.08$

The tension in the string is 94 N (2 s.f.).

Note: Use the value for *a* in ①.

Step 1: Find the final velocity of the particle while the string is taut.

c Find the velocity when the string breaks:

$$v^2 = u^2 + 2as \qquad \text{(where } s = 1, u = 0, v = ?, a = 3.5\text{)}$$

$$v^2 = 0^2 + 2 \times \frac{9g}{25} \times 1$$

$$v = \sqrt{7.056}$$

Note: *v* is the final velocity whilst the string is taut.

Step 2: Find the new acceleration of the particle when there is no tension.

Resolve the forces on *P*, parallel to the plane, after the string breaks:

$$-6g = 10a$$

$$a = -5.88$$

The acceleration of particle *P* after the string breaks is -5.88 m s^{-2}.

Recall: Use equation ① without the tension.

Step 3: Use the equations of motion to find the further distance moved by the particle until it stops.

Find the distance moved after the string breaks:

$$v^2 = u^2 + 2as \qquad \text{(where } u = \sqrt{7.056}, v = 0, s = ?, a = -5.88\text{)}$$

$$0^2 = 7.056 - 2 \times 5.88 \times s$$

$$s = 0.6$$

So the total distance travelled by *P* up the plane is 1.6 m.

Note: *a* is negative because the motion is decelerating.

SKILLS CHECK **6A: Connected particles**

1 Two particles of mass 5 kg and 8 kg are on a rough horizontal surface and are connected by a light, inextensible string. The 8 kg mass is being pulled by a horizontal force of 39 N. The resistance to each particle is *k* times their mass, so the resistance experienced by the 5 kg mass is 5*k* N. If the acceleration of the system is 1 m s^{-2}, find the tension in the connecting string and the magnitude of the resistance experienced by each particle.

 2 A dog of mass 6 kg is attached to a sleigh of mass 3 kg by a rope and pulls it along a smooth horizontal surface. The dog exerts a forward force of *F* N. The dog and the sleigh start from rest and travel a distance of 15 m in 10 seconds. Find the force with which the dog pulls the sleigh and find the tension in the rope. The dog and the sleigh now reach a smooth hill inclined at 1° to the horizontal and continue along the line of greatest slope of the plane. If the dog maintains the same pulling force find the acceleration of the dog and the sleigh up the plane. What assumptions have you made when modelling this situation?

3 A light, inextensible string connects two particles *A* and *B*. The particles hang vertically with particle *B* below particle *A*. Particle *A* has a mass of 3 kg and particle *B* has a mass of 2 kg. Particle *A* is pulled upwards by means of another string with tension 85 N. Find **a** the acceleration of the system and **b** the tension in the lower string.

4 A crane lowers two cars that are connected by a chain. The topmost car has mass 1000 kg and is suspended by a rope, which is connected to the crane. The lower car has mass 800 kg and is 100 m vertically above the ground.

    a Given that it takes 10 seconds for the lower car to reach the ground, starting from rest, find the tensions in **i** the chain and **ii** the rope.

    b What assumptions have you made about the chain and the rope?

5 Particles of mass 3 kg and 5 kg are attached by a light, inextensible string, which passes over a smooth fixed pulley. The 5 kg particle is 3 m above ground. The system is released from rest.

    a By finding the acceleration of the system, calculate how long it takes the 5 kg mass to reach the ground.

    b Also, find the force exerted on the pulley.

6 Particles P and Q are attached by a string that passes over a smooth fixed pulley. Particle P has twice the mass of particle Q. They both hang 2 m above horizontal ground. The system is released from rest.

    a **i** Find the magnitude of the acceleration of the system.
        ii Hence, find the velocity with which P hits the ground.

    b Assuming that particle Q does not reach the pulley, find the greatest height that Q reaches above the ground.

    c **i** How have you used the fact that the pulley is smooth?
        ii How have you used the fact that the string is light **and** inextensible?

7 Two particles A and B are connected by a light, inextensible string. Particle A has mass 6 kg and lies on a rough horizontal table. The string passes over a smooth fixed pulley at the edge of the table and particle B of mass 5 kg hangs vertically at the other end of the string, 2 m above horizontal ground. The system is released from rest and it takes 5 seconds for particle B to hit the ground. Find the acceleration of the system, then find the coefficient of friction between the plane and particle A.

 8 A particle P of mass 2.5 kg which is at rest on a smooth inclined plane of angle 30° is connected to particle Q of mass 3 kg by a light, inextensible string which lies along a line of greatest slope of the plane and passes over a fixed smooth pulley at the top of the plane. Particle Q hangs freely 1.5 m above horizontal ground. The system is released from rest with the string taut. Find:

    a the acceleration of the system

    b the tension in the string

    c the final velocity of Q when it hits the ground

    d the total distance that P moves up the plane, given that it does not reach the pulley.

9 The diagram shows two particles P and Q of masses 0.4 kg and 2 kg respectively. Particle P rests on a rough plane inclined at angle 30° to the horizontal and is attached to particle Q by means of a light, inextensible string which passes over a smooth fixed pulley at the top of the plane as shown. Particle Q lies on a smooth plane inclined at 60° to the horizontal. The coefficient of friction between particle P and the plane is $\frac{1}{\sqrt{3}}$. The system is released from rest.

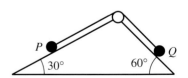

Find the acceleration of the system, given that Q travels down the plane.

SKILLS CHECK **6A EXTRA** is on the CD

1 Two particles of mass 2 kg and mass m kg are connected by a light inextensible string that passes over a smooth light pulley, as shown in the diagram below.

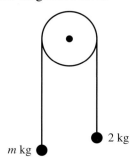

The particles are released from rest with the string taut.

a The 2 kg particle moves vertically downwards, travelling 0.8 metres in 2 seconds. Find its acceleration.

b Find the tension in the string.

c Find m, giving your answer to **two** significant figures.   [AQA (B) Jan 2003]

2

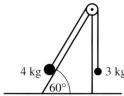

The diagram shows a car pulling a trailer in a straight line along a horizontal road. The car has mass 1000 kg and the resistance forces acting on the car are of magnitude 1200 N. The trailer has mass 250 kg and the resistance forces acting on the trailer are of magnitude 300 N. The forward propulsive force of the engine of the car is 2000 N.

a i By considering the car and the trailer as a single body, show that the acceleration of the car and the trailer is 0.4 m s^{-2}.

ii By considering the forces acting only on the trailer, or otherwise, find the tension in the coupling between the car and the trailer.

When the car and the trailer are moving with velocity 12 m s^{-1}, the coupling breaks and the trailer becomes separated from the car. During the subsequent motion the trailer moves under the resistance force only.

b i Show that the deceleration of the trailer during the subsequent motion is 1.2 m s^{-2}.

ii Find the distance moved by the trailer after the coupling breaks.   [AQA (A) Jan 2002]

 3 Two particles are connected by a light string that passes over a smooth, light pulley as shown in the diagram.

The 4 kg particle is on a smooth, fixed slope, which is at an angle of $60°$ to the horizontal. The 3 kg particle hangs with the string vertical.

The particles are released from rest at the position shown.

a Show that the acceleration of the particles is approximately 0.65 m s^{-2}.

b By considering the 3 kg particle, determine the tension in the string.   [AQA (B) Jan 2002]

4 [In this question the value of g should be taken to be $9.8 \, \text{m s}^{-2}$.]

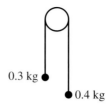

0.3 kg

0.4 kg

Two particles, of masses 0.3 kg and 0.4 kg, are connected by a light inextensible string which hangs over a smooth fixed peg, as shown in the diagram. The system is released from rest.

a i Show that, in the subsequent motion, the acceleration of the particles is of magnitude $1.4 \, \text{m s}^{-2}$.

ii Find the tension in the string during this motion.

b Find the distance travelled by each particle during the first 2 seconds of the motion.

[AQA (A) Jun 2001]

 5

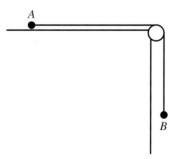

A

B

Two particles are connected by a light inextensible string, which passes over a smooth fixed peg, as shown in the diagram. The particle A, of mass 0.5 kg, is in contact with a rough horizontal surface, and the particle B, of mass 0.2 kg, hangs freely. The coefficient of friction between A and the surface is $\frac{2}{7}$.

The system is released from rest with the string taut and A moves towards the peg.

a Show that the frictional force between A and the surface is of magnitude 1.4 N.

b Find the acceleration of the particles.

c Find the tension in the string.

d Find the time taken for the particles to travel 0.625 metres, given that A has not then reached the peg. [AQA (A) Jun 2002]

6 Two particles, P and Q, are connected by a light inextensible string which passes over a smooth, fixed peg, as shown in the diagram.

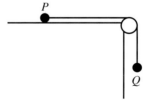

P

Q

The particle P is of mass $2m$ and Q is of mass m. The particle P is in contact with a horizontal surface, which may be modelled as either rough or smooth.

a In the first case, the surface is modelled as rough and the particle Q hangs at rest. The coefficient of friction between P and the surface is μ.

i Find the tension in the string.   **ii** Find the range of possible values of μ.

b In the second case, the surface is modelled as smooth. The system is released from rest with the string taut and the particle P moves towards the peg.

Find the tension in the string. [AQA (A) Jan 2003]

7 Projectiles

7.1 Projectiles

Motion of a particle under gravity in two dimensions.

In Chapter 2 you looked at the motion of a particle travelling in a straight line (horizontally or vertically). In this chapter we look at the motion of a particle that is free to move horizontally and vertically at the same time. This particle is called a **projectile**. The particle has both a horizontal motion and a vertical motion. When u m s^{-1} is the initial velocity and $\theta°$ is the angle that u makes with the horizontal, the path takes the following shape:

The aim is to apply the equations of motion horizontally and vertically to study the particle's path. The method is similar to the case of motion in a straight line. The following points are important to remember when filling the information into *suvat*.

1 The horizontal component of the motion does not experience any acceleration and so, horizontally, $a = 0$.

2 The vertical component of the motion travels freely under gravity and so, vertically, the acceleration is 9.8 m s^{-2} downwards.

3 When you are studying the path of a projectile, only time taken, t, will be the same for the horizontal and vertical motion. This is often the link between the equations of motion for each direction.

4 If a particle is projected with speed u m s^{-1} at an angle $\theta°$ above the horizontal, then you can work out the horizontal and vertical components of the velocity by resolving the velocity:

Recall:
Resolving forces into components in Chapter 2.

horizontal component of velocity $= u \cos \theta°$
vertical component of velocity $= u \sin \theta°$

At each moment in time the particle has a horizontal and a vertical component of velocity.
Remember to also take the direction into account by choosing a direction (upwards or downwards) as positive.

It is useful to review Section 2.4 before going on.

Horizontal projection

Here a particle is projected horizontally. In this case the initial speed vertically will be 0 m s^{-1}.

Example 7.1 A particle is projected horizontally with speed 8 m s^{-1} from the top of a cliff which is 80 m above the sea. Find the time taken for the particle to reach the surface of the sea and the horizontal distance travelled.

Step 1: Draw a clear diagram to represent the information given.	Taking downwards as positive:

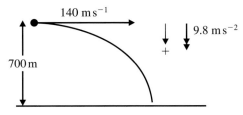

Horizontally:

$s = ?$

$u = 8$

v not required

$a = 0$

$s = ut + \frac{1}{2}at^2$

Vertically:

$s = 80$

$u = 0$

v not required

$a = 9.8$

$t = ?$

$s = ut + \frac{1}{2}at^2$

$80 = (0)t + \frac{1}{2}(9.8)t^2$

$t = 4.04\ldots$

Step 2: Fill the information in *suvat* (horizontally and vertically) identifying what is required with a question mark.

Step 3: Pick an equation of motion relating the required variable with the others. Insert values and solve.

Note:
You can see that we do not have enough information for the horizontal motion to find *s*, but you can use the time found from the vertical motion.

The time taken to reach the sea is 4.0 seconds (2 s.f.). This value can now be used to find the horizontal distance travelled.

$s = 8(4.04\ldots) + 0$

$\quad = 32.3\ldots$

The horizontal distance travelled is 32 m (2 s.f.).

Note:
Only the positive square root has been used for time.

Remember that the speed is the magnitude of the velocity. If you are given the (horizontal and vertical) components of the velocity, then you can find the speed by finding the resultant of the velocity.

Tip:
Because $a = 0$ horizontally, the equation $s = ut + \frac{1}{2}at^2$ becomes $s = ut$ for the horizontal motion.

Example 7.2 A plane which is travelling horizontally 700 m above the Earth's surface, with speed $140\ \mathrm{m\,s^{-1}}$, releases a package of food. In order for the container to land with the contents safe, the speed with which it hits the surface must be less than $200\ \mathrm{m\,s^{-1}}$. The package may be modelled as a particle.

a Find the horizontal and vertical components of the velocity when the package hits the ground.

b Determine whether or not the contents remain intact when the food package reaches the surface.

c Find the angle that the package makes with the horizontal on reaching the surface.

Step 1: Draw a clear diagram to represent the information given.

Taking downwards as positive:

Note:
It is assumed that the package will have the same speed and direction as the plane when it first leaves the plane.

Step 2: Fill the information in *suvat* (horizontally and vertically) identifying what is required with a question mark.

Step 3: Pick an equation of motion relating the required variable with the others. Insert values and solve.

a Horizontally:

s not required

$u = 140$

$v = ?$

$a = 0$

t not required

$\vdots$

$v = u$

$v = 140$

Vertically:

$s = 700$

$u = 0$

$v = ?$

$a = 9.8$

t not required

$\vdots$

$v^2 = u^2 + 2as$

$v^2 = 0^2 + 2(9.8)(700)$

$v = 117.13\ldots$

Tip:
Because $a = 0$ horizontally, the equation $v = u + at$ becomes $v = u$ for the horizontal motion.

b You now have the horizontal and vertical components of the velocity when the package lands. You can find the resultant using Pythagoras:

Final speed $= \sqrt{(140^2 + 117.13...^2)}$ m s^{-1}

$= 182.537...$ m s^{-1}

$= 180$ m s^{-1} (2 s.f.)

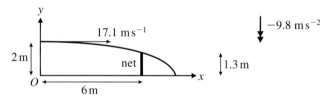

Recall:
Chapter 2 on finding the resultant of a vector quantity.

The final speed is less than 200 m s^{-1} so the package will land safely.

c Finally the angle, $\alpha°$, below the horizontal is given by

$$\tan \alpha° = \frac{117.13...}{140}$$

$$\alpha = 39.91...$$

The angle below the horizontal is $40°$ (2 s.f.).

Example 7.3 A tennis ball is served horizontally with speed 17.1 m s^{-1} from a point which is 2 m above the ground. The top of the net is 6 m away horizontally and has a height of 1.3 m.

a Find whether or not the tennis ball will pass over the net and, if it does, find the distance with which it clears the net.

b What assumptions have you made when modelling this situation?

Step 1: Draw a clear diagram to represent the information given.

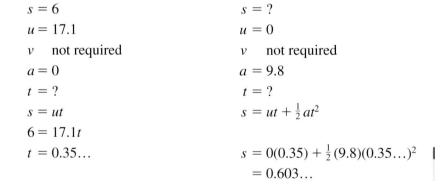

The idea is to find the vertical displacement of the tennis ball when it reaches the net – this is when the tennis ball has travelled 6 m horizontally.

Step 2: Fill the information in *suvat* (horizontally and vertically) identifying what is required with a question mark.

a horizontally:

$s = 6$

$u = 17.1$

v not required

$a = 0$

$t = ?$

vertically:

$s = ?$

$u = 0$

v not required

$a = 9.8$

$t = ?$

Step 3: Pick an equation of motion relating the required variable with the others. Insert values and solve.

$s = ut$

$6 = 17.1t$

$t = 0.35...$

$s = ut + \frac{1}{2}at^2$

$s = 0(0.35) + \frac{1}{2}(9.8)(0.35...)^2$

$= 0.603...$

Note:
Here, you use the time found from the horizontal motion and substitute into the vertical motion.

So, the displacement of the tennis ball vertically is 0.60 m (2 s.f.) downwards. Given that it started from a height of 2 m above the ground, the ball's distance from the ground when it reaches the net is $(2 - 0.603...)$ m $= 1.397...$ m. So it clears the net by $(1.397... - 1.3)$ m $= 0.967...$ m $= 0.10$ m (2 s.f.).

b The tennis ball has been treated as a particle, gravity has been assumed to be constant and air resistance and wind have been assumed to be negligible so that they can be ignored.

Projection at an angle to the horizontal

When a particle is projected at an angle to the horizontal then you can work out its horizontal and vertical components of the velocity (see point 4 at the start of the chapter).

Example 7.4   A particle is hit from a point O on the ground with speed 20 m s^{-1} at an angle of $30°$ above the horizontal. Assuming that the particle moves freely under gravity,

    a find the horizontal and vertical displacement from O after 1 second

    b hence, find the distance of the particle from O after 1 second.

Step 1: Draw a clear diagram to represent the information given.

Taking upwards as positive:

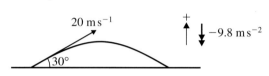

> **Note:**
> Vertically, $a = -9.8$ as it acts downwards and upwards motion has been taken as positive.

Step 2: Fill the information in *suvat* (horizontally and vertically) identifying what is required with a question mark.

a Horizontally:

$s = ?$

$u = 20 \cos 30°$

v not required

$a = 0$

$t = 1$

Vertically:

$s = ?$

$u = 20 \sin 30°$

v not required

$a = -9.8$

$t = 1$

Step 3: Pick an equation of motion relating the required variable with the others. Insert values and solve.

$s = ut$

$s = 20(\cos 30°)(1)$

$\;\;\;\; = 17.32\ldots$

$s = ut + \frac{1}{2}at^2$

$s = 20(\sin 30°)(1) - \frac{1}{2}(9.8)(1)^2$

$\;\;\;\; = 5.1$

After 1 second the horizontal displacement is 17 m (2 s.f.) and the vertical displacement is 5.1 m.

b You can now use Pythagoras' theorem to find the distance from the start:

$$\text{Distance} = \sqrt{(17.32\ldots^2 + 5.1^2)} \text{ m}$$
$$= 18.055\ldots \text{ m}$$
$$= 18 \text{ m (2 s.f.)}$$

> **Note:**
> Remember to use all the decimal places of the values that you put in here.

7.2 Range, time of flight and maximum height

Calculate range, time of flight and maximum height.

Time of flight and range

When a particle is projected from O on a surface and then comes back to the ground on the same level, the displacement vertically is 0 m from the start to the finish. So, vertically, $s = 0$. The time taken to come back to the surface again is called the **time of flight**. The horizontal distance travelled is called the **range**.

Example 7.5   A golf ball is hit from a tee towards a hole that is on the same level ground 75 m away with speed 25 m s^{-1} at an angle $\theta°$ to the horizontal where $\sin \theta° = \frac{3}{5}$.

   a   Given that the golf ball is free to fall under gravity, find the time for which the ball is in the air and how far the ball falls short of the hole.

   b   If there was a wind blowing in the horizontal direction of the motion of the golf ball, how would this affect the time of flight and the range of the golf ball?

Step 1: Draw a clear diagram to represent the information given.

Taking upwards as positive:

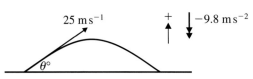

Recall:
Pythagorean triangles

$\sin \theta° = \frac{3}{5}$
$\cos \theta° = \frac{4}{5}$

Step 2: Fill the information in *suvat* (horizontally and vertically) identifying what is required with a question mark.

a   Horizontally:

$s = ?$
$u = 25 \cos \theta° = 20$
v not required
$a = 0$
$t = ?$

Vertically:

$s = 0$
$u = 25 \sin \theta° = 15$
v not required
$a = -9.8$
$t = ?$

Step 3: Pick an equation of motion relating the required variable with the others. Insert values and solve.

$s = ut$

$s = ut + \frac{1}{2}at^2$
$0 = 15t - 4.9t^2$
$= t(15 - 4.9t)$
$t = 0$ or $t = \frac{15}{4.9} = 3.06\ldots$

Note:
$t = 0$ refers to the start of the motion when $s = 0$ also.

Substituting $t = 3.06\ldots$ into the equation for horizontal motion gives:

$s = (20)(3.06\ldots) = 61.2\ldots$

Note:
The time of flight is 3.06 seconds.

So the time that the golf ball is in the air is 3.06 seconds (3 s.f.) and the range of its projectile is 61.2 m (3 s.f.). The ball falls short of the hole by

$(75 - 61.2\ldots) = 13.8\ldots$ m $= 13.8$ m (3 s.f.)

b   If there was a wind blowing in the direction of the ball, the time of flight would remain unchanged because this depends only on the vertical motion (which is unaffected by the direction of the wind). However, the range would increase because there would be a greater speed in the horizontal direction while in the air, due to the wind.

Maximum height

The **maximum height** reached by a particle occurs when the vertical component of the velocity $= 0$ m s^{-1}.

Example 7.6   A ball is projected from a point O with speed U m s^{-1} at an angle of $\theta°$ above the horizontal where

$$U \cos \theta° = 6$$
$$U \sin \theta° = 9.$$

The particle travels freely under gravity. Find:
   a   the maximum height above the origin that the particle reaches
   b   the horizontal displacement of the ball from the start at maximum height
   c   state the speed of the ball at the maximum height.

Recall:
$U \cos \theta°$ is the horizontal component of the velocity and $U \sin \theta°$ is the vertical component of the velocity.

Step 1: Draw a clear diagram to represent the information given.

Taking up as positive:

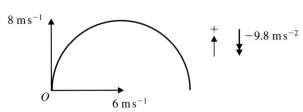

8 m s^{-1}

6 m s^{-1}

O

$+$ -9.8 m s^{-2}

Step 2: Fill the information in *suvat* (horizontally and vertically) identifying what is required with a question mark.

Horizontally:

s = ?

$u = 6$

v not required

$a = 0$

t = ?

Vertically:

s = ?

$u = 9$

$v = 0$

$a = -9.8$

t = ?

Step 3: Pick an equation of motion relating the required variable with the others. Insert values and solve.

a $v^2 = u^2 + 2as$

$0^2 = 9^2 + 2(-9.8)s$

$s = 4.13 \ldots$

So the maximum height reached by the particle is 4.1 m (2 s.f.).

b To find the horizontal displacement, first find the time taken from the vertical motion, then substitute into the appropriate equation for horizotnal motion.

$$v = u + at$$
$$0 = 9 - 9.8t$$
$$t = 0.918 \ldots$$

$s = ut$

$= (6)(0.918\ldots) = 5.51\ldots$

The horizontal displacement at maximum height is 5.5 m (2 s.f.).

c The speed of the particle at the maximum height is the horizontal component of the speed because the vertical component = 0 m s^{-1} at the maximum height. Horizontally, the speed remains unchanged from the start as there is no acceleration, so the speed of the particle at maximum height = 6 m s^{-1}.

Recall:
$v = u$ horizontally.

The path of a projectile is symmetrical about the maximum height. So in the above question the range of the particle is twice the horizontal distance travelled from the starting point (11 m), and the time of flight of the particle is twice the time taken to reach the maximum height (1.8 seconds).

7.3 Modification of the equations of motion

Modification of the equations of motion to take account of the height of release.

When a particle is projected from a point above ground, the displacement is measured from the starting point. The distance above the ground from which the particle starts must also be taken into account.

Example 7.7 A ball is projected with speed 50 m s⁻¹ at an angle of elevation of 30° from the top of a cliff that is 100 m above sea level. The ball travels freely under gravity. Find:

a the maximum height above sea level reached by the particle

b the time taken for the particle to reach the sea

c the horizontal distance of the particle from the edge of the cliff when it lands in the sea.

Note:
An **angle of elevation** is the angle measured above the horizontal.

Step 1: Draw a clear diagram to represent the information given.

Take upwards as positive and let h be the maximum height reached above the starting point on the cliff. First analyse the motion from the start point to the maximum height.

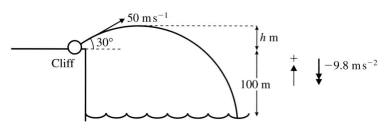

a Vertically:

$$s = h$$
$$u = 50 \sin 30°$$
$$v = 0$$
$$a = -9.8$$
$$t \quad \text{not required}$$
$$v^2 = u^2 + 2as$$
$$0^2 = (50 \sin 30°)^2 - 19.6h$$
$$h = 31.888\ldots$$

Step 2: Fill the information in *suvat* (horizontally and vertically) identifying what is required with a question mark.

Step 3: Pick an equation of motion relating the required variable with the others. Insert values and solve.

The height above sea level is $(100 + 31.8\ldots) = 130$ m (2 s.f.).

For the next two parts you need to analyse the motion from where the particle starts to the point where it hits the sea. So you need to fill in a new set of information in *suvat* as there is a different motion. The diagram is the same, so start with Step 2.

Step 2: Fill the information in *suvat* (horizontally and vertically) identifying what is required with a question mark.

Horizontally:
$$s = ?$$
$$u = 50 \cos 30°$$
$$v \quad \text{not required}$$
$$a = 0$$
$$t = ?$$

Vertically:
$$s = -100$$
$$u = 50 \sin 30°$$
$$v \quad \text{not required}$$
$$a = 9.8$$
$$t = ?$$

Step 3: Pick an equation of motion relating the required variable with the others. Insert values and solve.

b $\quad s = ut + \frac{1}{2}at^2$
$$-100 = (50 \sin 30°)t - 4.9t^2$$
$$4.9t^2 - 25t - 100 = 0$$
$$t = -2.64\ldots \text{ or } t = 7.74\ldots$$

The particle takes 7.7 s (2 s.f.) to reach the sea.

c $s = ut$
$$= (50 \cos 30°)(7.74\ldots)$$
$$= 335.11\ldots$$

So the horizontal distance travelled is 340 m (2 s.f.).

Recall:
s is measured from the start point which is at the top of the cliff.

Tip:
With practice you will notice that sometimes the vertical motion is enough.

Tip:
It is useful to remember that $2 \times 9.8 = 19.6$ and $\frac{9.8}{2} = 4.9$ for ease in calculation.

Note:
$t = -2.64\ldots$ is not physically reasonable and so is rejected.

Example 7.8 A ball is thrown out of a window that is 60 m above the ground with velocity U, at an angle of $\theta°$ *below* the horizontal where $\tan \theta° = \frac{5}{12}$. The ball just clears a fence that is at a horizontal distance of 48 m from the window. The fence is 20.4 m high. Let T be the time taken to reach the fence.

Note:
This angle can also be described as an **angle of depression**.

a Show that $UT = 52$.

Step 1: Draw a clear diagram to represent the information given.

b Find the values of U and T.

Take downwards as positive.

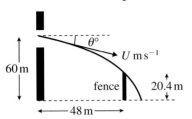

Step 2: Fill the information in *suvat* (horizontally and vertically) identifying what is required with a question mark.

a Horizontally:
$s = 48$
$u = U \cos \theta° = \dfrac{12U}{13}$
v not required
$a = 0$
$t = T$
$s = ut$

Step 3: Pick an equation of motion relating the required variable with the others. Insert values and solve.

$48 = \dfrac{12UT}{13}$
$UT = 52$

Substitute 52 for UT into the equation for vertical motion.

b Vertically:
$s = 60 - 20.4 = 39.6$
$u = U \sin \theta° = \dfrac{5U}{13}$
v not required
$a = 9.8$
$t = T$
$s = ut + \frac{1}{2}at^2$
$39.6 = \dfrac{5UT}{13} + 4.9T^2$
$39.6 = \dfrac{5(52)}{13} + 4.9T^2$
$T = 2$

Therefore $U = \dfrac{52}{T} = 26$

Recall:
Pythagorean triangles

$\sin \theta° = \frac{5}{13}$
$\cos \theta° = \frac{12}{13}$

Note:
Each equation does not have sufficient information, but you can solve simultaneously for the unknowns.

The initial speed of the particle is 26 m s^{-1} and the time taken to reach the fence is 2 seconds.

SKILLS CHECK **7A: Projectiles**

1 A particle is projected horizontally with speed 8 m s^{-1}, from a point that is 40 m above ground, and travels freely under gravity.
  a Find **i** the time taken to reach the ground and **ii** the horizontal distance travelled in this time.
  b What is the speed of the ball immediately before it hits the ground?

2 An archer fires an arrow horizontally with speed 250 m s^{-1} and the arrow is free to move through the air under gravity. The arrow is fired from a height of 1.8 m above ground. Find:
  a how long it takes for the arrow to hit the ground
  b the horizontal displacement from the start when it does hit the ground.

3 A ball is fired horizontally from a cannon that lies on the top of a tower. The foot of the tower is a horizontal distance of 100 m from where the ball lands on the ground and it takes 4 seconds to reach the point where it lands. Assuming that the ball is free to move under gravity, find:
  a the speed of projection
  b the height of the tower
  c the speed of the particle two seconds after it has been fired.

4 A particle is is projected horizontally with speed U m s^{-1} from a point which is 9 m above the ground. The particle travels freely under gravity. After T seconds it has reached the ground and has travelled 6 m in the horizontal direction. Find T and U.

5 A particle is projected from a point O with speed U m s^{-1} at an angle $\theta°$ above the horizontal where

$$U \cos u° = 2$$
$$U \sin u° = 3.$$

The particle travels freely under gravity. Find the speed after it has travelled for 3 seconds.

6 A gun fires a bullet from ground with speed 120 m s^{-1} at an angle of 70° above the horizontal. By modelling the bullet as a particle that moves freely under gravity, find:

a the maximum height reached

b the time taken to reach the maximum height

c the horizontal distance travelled by the bullet when it reaches its maximum height

d the time of flight (the time taken for the bullet to reach the ground again) and the range of the bullet.

e What assumptions have you made when modelling this situation?

 7

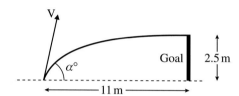

The diagram represents a football player, who is about to take a penalty using a home-made goal. He wants to aim the ball straight ahead at the goalkeeper, with the intention of the ball reaching its maximum height just under the crossbar, which is 2.5 m high. The initial speed of the ball is V m s^{-1} and it is kicked at an angle of elevation of $\alpha°$ above the ground. The penalty spot is 11 m from the goal.

a Show that $V \sin \alpha° = 7$.

b Find the time taken for the ball to reach the goal.

c Find α and V.

 8 For a particle that moves freely under gravity with initial speed U m s^{-1}, travelling at an angle of $\theta°$ above the horizontal, show that the maximum height, h m, reached above the starting point is given by

$$h = \frac{U^2 \sin^2 \theta}{2g}$$

where g is the acceleration due to gravity.

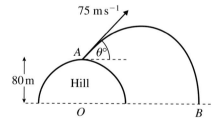

A golf ball is hit from a point A on the top of a hill with speed 75m s^{-1} at an angle of elevation $\theta°$, where $\sin \theta = \frac{4}{5}$. The height of the hill, OA, is 80 m above the horizontal ground where O is the foot of the perpendicular below A. The ball lands on the ground at the point B. Assuming that the golf ball travels freely under gravity, find:

a the maximum height above the ground that the golf ball reaches

b the time taken for the golf ball to reach the ground

c the distance OB

d the direction in which the golf ball is moving when it hits the ground.

9 A particle is projected from 150 m above ground with speed V m s^{-1} at an angle of elevation of $\theta°$ and travels freely under gravity. After 2 seconds, the horizontal component of the velocity is $(10\sqrt{5})$ m s^{-1} and the vertical component of the velocity is $(19.6 - 5\sqrt{5})$ m s^{-1} downwards.

  a Show that $\tan \theta = \frac{1}{2}$.

  b Find the initial speed of the particle.

  c Find the maximum height above ground reached by the particle.

  d Find the time taken for the particle to reach the ground.

SKILLS CHECK **7A EXTRA** is on the CD

Examination practice Projectiles

1

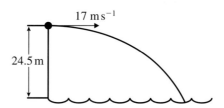

A child throws a stone from the top of a vertical cliff and the stone subsequently lands in the sea. The stone is thrown from a height of 24.5 metres above the level of the sea. The initial velocity of the stone is horizontal and has magnitude 17 m s^{-1}, as shown in the diagram.

  a Find the time between the stone being thrown and reaching the sea.

  b Find the horizontal distance between the foot of the cliff and the point where the stone reaches the sea.

  c Find the **speed** of the stone as it reaches the sea.   [AQA (A) Nov 2002]

2 The diagram shows a target that is used for rifle shooting. It consists of three concentric circles of radii 15 cm, 25 cm and 40 cm respectively.

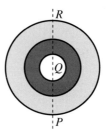

When the rifle is fired it is always 50 metres from the target and at the same height as the centre of the target. The bullet moves in the vertical plane that continues the line PR. The bullet is assumed not to be subject to any air resistance.

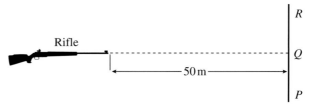

A bullet is fired horizontally. It hits the target on the outer circle at the point P. Show that the bullet was fired at a speed of 175 m s^{-1}.   [AQA (B) Jan 2002]

3 A particle is projected from a horizontal surface at a speed V and at an angle α above the horizontal.

    a Prove that the maximum height of the particle is $\dfrac{V^2 \sin^2 \alpha}{2g}$.

    b A ball is hit from ground level. The ball initially moves at an angle of $60°$ above the horizontal. The maximum height of the ball is 6 metres above the ground. Modelling the ball as a particle:

       i find the initial speed of the ball;

      ii find the range of the ball.   [AQA (B) Jan 2003]

 4 A javelin is modelled as a particle. Assume that only gravity acts on the javelin after it has left the thrower's hand. The initial velocity of the javelin is 20 m s^{-1} at an angle of $40°$ above the horizontal.

    a Find the range of the javelin on horizontal ground if the height of release is ignored.

    b The javelin is actually released at a height of 2 metres. Find the range of the javelin in this case.   [AQA (B) Jun 2001]

5 A golfer hits a ball, from ground level on a horizontal surface. The initial velocity of the ball is 21 m s^{-1} at an angle of $60°$ above the horizontal. Assume that the ball is a particle and that no resistance forces act on the ball.

    a Find the maximum height of the ball.

    b Find the range of the ball.

    c Find the speed of the ball at its maximum height.   [AQA (B) Jan 2001]

Practice exam paper

Answer **all** questions.

Time allowed: 1 hour 20 minutes

A calculator **may** be used in this paper.

1 A ball is thrown vertically upwards from ground level with speed u m s^{-1} and takes 4 s to reach the ground again. Neglecting air resistance,

   a draw a velocity–time graph to represent the motion of the ball during the first 4 s   *(2 marks)*

   b find the maximum height of the ball above the ground.   *(3 marks)*

2 A particle P is suspended from two light inextensible strings and hangs in equilibrium. One string is inclined at $60°$ to the horizontal and the tension in that string is 40 newtons. The other string is inclined at $30°$ to the horizontal. Find

   a the weight of P *(3 marks)*

   b the tension in the other string.   *(3 marks)*

3 A particle is acted upon by two forces $\mathbf{F_1}$ and $\mathbf{F_2}$.

$$\mathbf{F_1} = \begin{bmatrix} 3 \\ 1 \end{bmatrix} \text{ newtons and } \mathbf{F_2} = \begin{bmatrix} a \\ 2a \end{bmatrix} \text{ newtons, where } a \text{ is a constant.}$$

   a Find the angle between $\mathbf{F_1}$ and $\begin{bmatrix} 0 \\ 1 \end{bmatrix}$.   *(2 marks)*

The resultant $\mathbf{R}$ of $\mathbf{F_1}$ and $\mathbf{F_2}$ is parallel to $\begin{bmatrix} 1 \\ 0 \end{bmatrix}$.

   b Find the magnitude of R.   *(6 marks)*

4 A particle P moves so that its velocity v m s^{-1} at time t seconds is given by $v = (3t^2 + 1)\mathbf{i} + (7t - 1)\mathbf{j}$.

   a Find the initial speed of P.   *(2 marks)*

   b At what time is P first moving parallel to the vector $\begin{bmatrix} 1 \\ 1 \end{bmatrix}$?   *(4 marks)*

5 A children's slide is straight and inclined at $30°$ to the horizontal. A child of mass 30 kg slides from rest down the slide.

   a In a **first** model, the slide is assumed to be perfectly smooth. Find the speed of the child after sliding a distance of 20 m down the slide.   *(4 marks)*

   b In a **second** model the slide is assumed to be rough with a coefficient of friction of 0.5. Find the speed of the child after sliding a distance of 20 m down the slide.   *(6 marks)*

6 The diagram below shows a light inextensible string passing over a smooth fixed pulley. Attached to one end of the string is a particle A of mass $2m$ and attached to the other end are particles B and C, of masses m and $2m$ respectively, which are joined together.

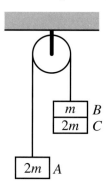

The system is released from rest.

a Show that the acceleration of the masses in the subsequent motion is 1.96 m s^{-2}. *(5 marks)*

b Find the speed of A after it has travelled 50 cm. *(2 marks)*

After C has travelled 50 cm it falls off and the system continues without it. Assuming that A does not reach the pulley,

c find how much further B moves down before it next comes to rest. *(6 marks)*

7 During a game of catch a ball is thrown from a point 1 m above horizontal ground. The ball is projected at an angle α above the horizontal, where $\tan \alpha = \frac{3}{4}$. The ball hits the ground at a point which is a horizontal distance 2 m from its point of projection, as shown in the diagram. The initial speed of the ball is u m s^{-1} and the time of flight is T seconds.

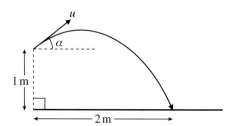

a Show that $2uT = 5$. *(2 marks)*

b Find the value of u. *(5 marks)*

c Find the direction of motion of the ball as it hits the ground. *(5 marks)*

Answers

SKILLS CHECK 2A (page 10)

1 a speed (m s⁻¹)

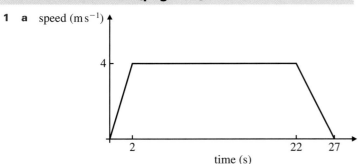

b $2\,\text{m s}^{-2}$, $-\frac{4}{5}\,\text{m s}^{-2}$ **c** 94 m

2 a velocity (m s⁻¹)

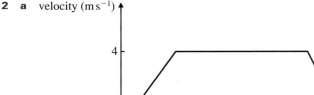

b 7 s   **c** 126 m

d acceleration (m s⁻²)

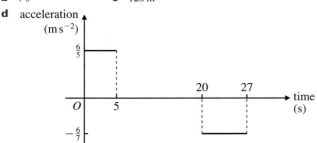

e $4\frac{2}{3}\,\text{m s}^{-1}$

3 a

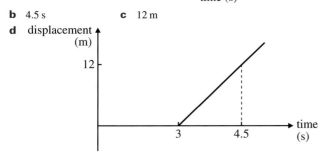

b 4.5 s   **c** 12 m

d displacement (m)

4 a $2\,\text{m s}^{-2}$ **b** 4 m
5 $\frac{25}{18}\,\text{m s}^{-2}$
6 $9\,\text{m s}^{-1}$
7 a $\frac{2}{3}\,\text{s}$ **b** No air resistance, car is a particle
8 50 m
9 a 13.9 s   **b** 965 m
10 a $2.5\,\text{m s}^{-2}$ **b** 140 m
11 a 4 m   **b** 4.674 s   **c** 10.8 m

SKILLS CHECK 2B (page 14)

1 $8.4\,\text{m s}^{-1}$, 36.4 m
2 40 m, 5.7 s, 4.04 s
3 4.04 s; book is a particle, no air resistance, gravity constant, book starts from rest
4 a 14 m   **b** 3.12 s
5 $31.3\,\text{m s}^{-1}$
6 a 5.63 s   **b** 155.3 m
7 b 6.38 m

SKILLS CHECK 2C (page 18)

1 a 5 km, 307°   **b** 5.73 km, 282°   **c** 4.53 km, 314°
2 a 5, 53.1°   **b** 6.08, 351° (or −9.46°)
   c 2.83, 225°   **d** 6.71, 117°
3 a $8.46\mathbf{i} + 3.08\mathbf{j}$ **b** $-3.94\mathbf{i} + 0.69\mathbf{j}$ **c** $-12.1\mathbf{i} - 7\mathbf{j}$
4 a $5\mathbf{i} - 3\mathbf{j}$, 5.83, 121°   **b** $\mathbf{i} + 5\mathbf{j}$, 5.10, 11.3°   **c** $8\mathbf{i} - 2\mathbf{j}$, 8.25, 104°
   d $-5\mathbf{i} - 11\mathbf{j}$, 12.1, 156°   **e** $4\mathbf{i} + 6\mathbf{j}$, 7.21, 33.7°

SKILLS CHECK 2D (page 24)

1 a $\mathbf{i} + \mathbf{j}$, $1.41\,\text{m s}^{-1}$ **b** $-\mathbf{i} + 2\mathbf{j}$, $2.24\,\text{m s}^{-1}$ **c** $1.5\mathbf{i} + 5\mathbf{j}$, $5.22\,\text{m s}^{-1}$
2 a $(2l\mathbf{i} + 24\mathbf{j})\,\text{m}$ **b** $(24\mathbf{i} + 25\mathbf{j})\,\text{m}$ **c** 34.7 m, 46.2°
3 a $(1.5\mathbf{i} + 4.5\mathbf{j})\,\text{m s}^{-2}$ **b** $(16\mathbf{i} - 16\mathbf{j})\,\text{m s}^{-2}$ **c** $\mathbf{j}\,\text{m s}^{-2}$
4 $20\,\text{m s}^{-1}$
5 a $[(3\mathbf{i} + 8\mathbf{j}) + t(\mathbf{i} - 2\mathbf{j})]\,\text{m} = [(3 + t)\mathbf{i} + (8 - 2t)\mathbf{j}]\,\text{m}$
   b $(6\mathbf{i} + 2\mathbf{j})\,\text{m}$ **c** 2
6 a $2\mathbf{i}$, $4\mathbf{j}$ **b** $2.2\,\text{m s}^{-1}$ **c** $(2\mathbf{i} + 4\mathbf{j})\,\text{m}$
7 a $(\mathbf{i} - \mathbf{j})\,\text{m s}^{-2}$ **b** $(70\mathbf{i})\,\text{m}$ **c** 70 m
8 a 34 m   **b** i 3 s   ii 2 s
9 a $(3\mathbf{i} - 8\mathbf{j})$, $8.54\,\text{m s}^{-1}$, 159°   **b** $-2, 8$
10 a $13\,\text{m s}^{-1}$, 22.6°   **b** $10.9\,\text{m s}^{-1}$, 65.4°
11 $40.2\,\text{m s}^{-1}$, 014.4°

Exam practice 2 (page 25)

1 a

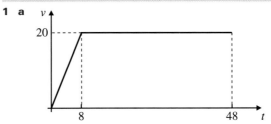

b 880 m

2 a

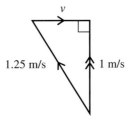

   b $0.75\,\text{m s}^{-1}$
   c 53.1°

3 a $3\mathbf{i}$ **b** $3\mathbf{i} + (0.1\mathbf{i} + 0.2\mathbf{j})t$
   c $0.2T = 0.1T + 3$ **d** 89.4 m
4 a $1.2\mathbf{i} + 0.8\mathbf{j}$ **b** $405\mathbf{i} + 90\mathbf{j}$
5 a 2.5 m   **b** 1.43

SKILLS CHECK 3A (page 31)

1 a 6.7 N, 63.4° b 7.1 N, 135.0° c 20.6 N, 29.1°
   d 3.6 N, 33.7° e 5.8 N, 301.0° (or −59.0°)
2 a 9.8 N, 1.7 N b 24.1 N, 6.5 N c 3.1 N, 34.9 N
3 a −2.5**i** + 4.3**j** b 17.3**i** − 10.0**j** c −111.1**i** − 93.2**j**
4 a 16.2 N, 4.0° b 98.5 N, 8.54°
5 a 15.1 N, 7.59° b 5.39 N, 248°
6 a 10 N, 8.45 N b 18.8 N, 12.3 N
7 11.7 N, 190°
8 a −14, −4 b 1, 10 c 2, 14

SKILLS CHECK 3B (page 38)

1 a 1.96 N, 7 N b 29.4 N, 14 N c 88.2 N, 24 N
2 a 10.4 N, 43 N b 13.9 N, 0.91 kg c 6.93 N, 5.36 N
3 a 150 N b 0.01
4 a 24.8 N, 0.78 b 1.36 N, 5.25 N c 0.97, 1.95 kg
5 73.6 N, 69.1 N
6 a 12.3 N b 1.61 kg
7 a i $\frac{12}{13}mg$ ii $\frac{5}{13}mg$ b $\frac{5}{12}$
8 a 27.4 N b 1.97 N
9 a 44.5 N b 5.57 N
10 a 5180 N b 13 300 N

Exam practice 3 (page 39)

2 b 283
3 a 9**i** − 4**j** b 9.85 N c 24.0°
4 a $\begin{bmatrix} 21.7 \\ 12.5 \end{bmatrix}$ b $\begin{bmatrix} -7.5 \\ -26.6 \end{bmatrix}$
5 a At least 3 of: weight, vertical; reaction, vertical; tension, oblique; friction, horizontal and opposite to tension.

   b 48 N d 200 N

SKILLS CHECK 4A (page 44)

1 a 8 N s b 1.8 N s c 10 N s
2 a 96 N s b 128 N s
3 a 4.67 m s⁻¹ b 0.75 m s⁻¹ c 26 m s⁻¹
4 3.69 m s⁻¹
5 a 300 m s⁻¹ b 45 000 N
6 a $\begin{bmatrix} 4.5 \\ 1.5 \end{bmatrix}$ b 4.74 m s⁻¹
7 a (3**i** + 4**j**) b i 5 m s⁻¹ ii 53.1°

Exam practice 4 (page 45)

1 0.4
2 b 1.71 m s⁻¹
   c Assumptions: no air resistance; bullet and block are particles; gun is free to recoil
3 a $\begin{bmatrix} -3 \\ 4 \end{bmatrix}$ b 5 m s⁻¹ c B moves faster.
4 a 0.025 b 5
5 b 1.5 m s⁻¹ c i −0.5 m s⁻² ii 2.25 m
6 b 18.6

SKILLS CHECK 5A (page 51)

1 a 4 m s⁻² b 2 m s⁻²
   c 5 m s⁻² d 0.56 m s⁻²
2 a 6 N b 1.65 N c 0.27 N
3 a 2.35 m s⁻² b 18.8 m

4 4500 N
5 4273 N
6 a 39.9 kg
   b Wall treated as a particle; neglect air resistance; rope is light and inextensible.
7 6.9 m s⁻²
8 8780 N
9 a $\frac{5}{12}$ b 1.62 m s⁻², 3.23 m
10 a 3.16 m s⁻² b −34.7°
11 3.5, −2

Exam practice 5 (page 52)

1 a ii 0.54 (or 0.55) m s⁻² b 114 N
2 a i

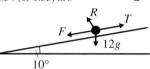

   ii 116 N

   b 1.20 m s⁻²
   c Sledge is a particle
      No air resistance
3 a i 6 m s⁻¹ ii −12 m s⁻²
   c 7.6 m s⁻² d 1.13 s
4 a

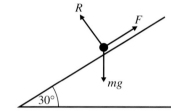

   c 3.2 m s⁻²
   d 2.77 m s⁻¹
5 a −10**i** − 5**j** b 19.9 N
6 a ii 2.8 m s⁻² b i 69.5 N ii 5.3 m s⁻¹

SKILLS CHECK 6A (page 58)

1 15 N, 10 N, 16 N
2 2.7 N, 0.9 N; 0.13 m s⁻²; the rope is light and inextensible, and there is no air resistance.
3 a 7.2 m s⁻² b 34 N
4 a i 6240 N ii 14 040 N
   b The chain is light and the rope is light and inextensible.
5 a 2.45 m s⁻² b 1.56 s, 73.5 N
6 a i 3.27 m s⁻² ii 3.61 m s⁻¹ b 4.67 m
   c i Tensions are the same in the string either side of the pulley.
      ii There is no component weight for the string; there is no force used in extending the string.
7 0.803
8 a 3.12 m s⁻² b 20.0 N
   c 3.06 m s⁻¹ d 2.46 m
9 6.91 m s⁻²

Exam practice 6 (page 60)

1 a 0.4 m s⁻² b 18.8 N c 1.8
2 a ii 400 N b ii 60 m
3 b 31.3 N
4 a ii 3.36 N b 2.8 m
5 b 0.8 c 1.8 N d 1.25 s
6 a i $T = mg$ ii $\mu \geqslant \frac{1}{2}$ b $T = \frac{2mg}{3}$

SKILLS CHECK 7A (page 69)

1 a i 2.86 s ii 22.9 m b 29.1 m s⁻¹
2 a 0.61 s b 152 m
3 a 25 m s⁻¹ b 78 m c 32 m s⁻¹

4   1.36 s, 4.43 m s^{-1}
5   26.5 m s^{-1}
6   **a**   649 m   **b**   11.5 s   **c**   472 m   **d**   23.0 s, 945 m
    e   Gravity constant, no air resistance, no wind, bullet treated as particle
7   **b**   0.71 s   **c**   24.4, 16.9 m s^{-1}
8   **a**   264 m   **b**   13.5 s   **c**   606 m
    d   At an angle of 58.0° to the ground
9   **b**   25 m s^{-1} **c**   156 m   **d**   6.80 s

Exam practice 7 (page 71)

1   **a**   2.24 s   **b**   38.0 m   **c**   27.7 m s^{-1}
3   **b**   **i** 12.5 m s^{-1} **ii** 13.9 m
4   **a**   40.2 m   **b**   42.5 m
5   **a**   16.9 m   **b**   39.0 m   **c**   10.5 m s^{-1}

Practice exam paper (page 73)

1   **a**

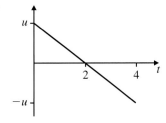

    b   19.6 m
2   **a**   46.2 N (3 s.f.)   **b**   23.1 N (3 s.f.)
3   **a**   71.6° (3 s.f.)   **b** $|R| = 2.5$ N
4   **a**   1.4 m s^{-1} **b** $\frac{1}{3}$ second
5   **a**   14 m s^{-1} **b**   5.12 m s^{-1} (3 s.f.)
6   **b**   1.4 m s^{-1} **c**   0.3 m
7   **b**   3.5   **c**   60.3° to the horizontal.

SINGLE USER LICENCE AGREEMENT FOR MECHANICS 1 FOR AQA CD-ROM
IMPORTANT: READ CAREFULLY